EVALUATING CLIMATE CHANGE ACTION PLANS

National Actions for International Commitment

ENVIRONMENTAL SCIENCE RESEARCH

Series Editor:

Herbert S. Rosenkranz
Department of Environmental and Occupational Health
Graduate School of Public Health
University of Pittsburgh
130 DeSoto Street
Pittsburgh, Pennsylvania

Founding Editor:

Alexander Hollaender

Recent Volumes in this Series

Volume 43 — PRIMARY PRODUCTIVITY AND BIOGEOCHEMICAL CYCLES IN THE SEA
Edited by Paul G. Falkowski and Avril D. Woodhead

Volume 44 — SECONDARY-METABOLITE BIOSYNTHESIS AND METABOLISM
Edited by Richard J. Petroski and Susan P. McCormick

Volume 45 — GLOBAL CLIMATE CHANGE: Linking Energy, Environment, Economy, and Equity
Edited by James C. White

Volume 46 — PRODUCED WATER: Technological/Environmental Issues and Solutions
Edited by James P. Ray and F. Rainer Engelhardt

Volume 47 — GLOBAL ENERGY STRATEGIES: Living with Restricted Greenhouse Gas Emissions
Edited by James C. White

Volume 48 — GLOBAL ATMOSPHERIC–BIOSPHERIC CHEMISTRY
Ronald G. Prinn

Volume 49 — BIODEGRADATION OF NITROAROMATIC COMPOUNDS
Edited by Jim C. Spain

Volume 50 — BIOMONITORS AND BIOMARKERS AS INDICATORS OF ENVIRONMENTAL CHANGE: A Handbook
Edited by Frank M. Butterworth, Lynda D. Corkum, and Judith Guzmán-Rincón

Volume 51 — CHEMISTRY FOR THE PROTECTION OF THE ENVIRONMENT 2
Edited by Lucjan Pawłowski, William J. Lacy, Christopher G. Uchrin, and Marzenna R. Dudzińska

Volume 52 — PRODUCED WATER 2: Environmental Issues and Mitigation Technologies
Edited by Mark Reed and Ståle Johnsen

Volume 53 — EVALUATING CLIMATE CHANGE ACTION PLANS: National Actions for International Commitment
Edited by James C. White

A Continuation Order Plan is available for this series. A continuation order will bring delivery of each new volume immediately upon publication. Volumes are billed only upon actual shipment. For further information please contact the publisher.

EVALUATING CLIMATE CHANGE ACTION PLANS

National Actions for International Commitment

Edited by

James C. White

Cornell University
Ithaca, New York

Associate Editors:

Wendy H. Petry and
William R. Wagner

Center for Environmental Information
Rochester, New York

Plenum Press • New York and London

Library of Congress Cataloging-in-Publication Data

On file

Proceedings of the Tenth Annual Conference of the Center for Environmental Information, Inc., on National Actions for International Commitment: Evaluating Climate Change Action Plans, held November 30 – December 2, 1994, in Washington, D.C.

ISBN-13: 978-1-4613-8006-1 e-ISBN-13; 978-1-4613-0341-1
DOI: 10.1007/978-1-4613-0341-1

Softcover reprint of the hardcover 1st edition 1997

A Division of Plenum Publishing Corporation
233 Spring Street, New York, N.Y. 10013

10 9 8 7 6 5 4 3 2 1

ABOUT THE CENTER FOR ENVIRONMENTAL INFORMATION

The Center for Environmental Information (CEI) is a private, nonprofit, educational organization, founded in Rochester, New York, in 1974. CEI provides information and communication services, publications, and educational programs in order to:

- advance public understanding of environmental issues;
- act as a communication link among scientists, educators, decision makers and the public;
- advocate informed action based on the free exchange of information and ideas.

CONFERENCE STEERING COMMITTEE

Donna Mercado Banks, Manager, Environmental Analysis, American Gas Association
J. Christopher Bernabo, President, Science & Policy Associates, Inc.
Alex Cristofaro, Director, Air and Energy Policy Division, U.S. Environmental Protection Agency
Roger C. Dower, Director, Climate, Energy and Pollution Program, World Resources Institute
Hadi Dowlatabadi, Department of Energy and Public Policy, Carnegie-Mellon University
Jae Edmonds, Battelle Pacific Northwest Laboratories
Howard Gruenspecht, Associate Deputy Undersecretary for Program Analysis, U.S. Department of Energy
Lester Machta, Air Resources Laboratory, National Oceanic and Atmospheric Administration
David Montgomery, Charles River Associates
Richard H. Moss, Intergovernmental Panel on Climate Change
Richard Richels, Director, Energy Analysis and Planning Department, Electric Power Research Institute
Martin A. Smith, Chief Environmental Scientist, Niagara Mohawk Power Corporation
Carol Werner, Director, Energy Program, Environmental and Energy Study Institute
James C. White, Center for the Environment, Cornell University
Thomas Wigley, UCAR Office for Interdisciplinary Earth Studies
Richard Wilson, Department of Physics, Harvard University

Conference Cosponsors

Contributing Cosponsors

- American Automobile Manufacturers Association
- American Gas Association
- Edison Electric Institute
- Electric Power Research Institute
- Niagara Mohawk Power Corporation
- U.S. Department of Agriculture, Economic Research Service
- U.S. Department of Energy
- U.S. Environmental Protection Agency, Air and Energy Policy Division
- World Resources Institute

Cooperating Cosponsors

- Air & Waste Management Association
- Alliance to Save Energy
- American Association for the Advancement of Science
- American Council for an Energy Efficient Economy
- Climate Institute
- Cornell University, Center for the Environment
- Environment Canada
- Environmental and Energy Study Institute
- Environmental Defense Fund
- Friends of the Earth
- Global Climate Change Digest
- Global Climate Coalition
- National Council of the Paper Industry for Air and Stream Improvement, Inc.
- National Rural Electric Cooperative Association
- Natural Resources Defense Council
- Princeton University, Center for Energy & Environmental Studies
- Smithsonian Institution
- Society of American Foresters
- Solar Energy Industries Association
- State & Territorial Air Pollution Program Administrators and The Association of Local Air Pollution Control Officials
- Union of Concerned Scientists
- U.S. Agency for International Development
- U.S. Department of Agriculture, Global Change Program
- University of Maryland at College Park, Center for Global Change
- World Meteorological Organization
- World Wildlife Fund

PREFACE

The atmospheric scientists of the world are in general agreement that the threat of climate change is real, inevitable, and serious. The accumulation of greenhouse gases, principally CO_2 from burning fossil fuels, is the main cause.

At the 1992 United Nations Conference on Environment and Development in Rio de Janeiro, 166 nations signed the Framework Convention on Climate Change and agreed to draw up plans to contain greenhouse gases at 1990 levels. Never in world history had so many nations agreed on anything.

Developing these plans has not been easy and no two countries have had the same circumstances and conditions to meet. Countries have not approached their problems in the same manner and many nations find the task almost impossible to solve under the ground rules set up in Rio.

This volume contains the papers presented at a meeting organized by the Center for Environmental Information and held in Washington, D.C., November 30 to December 2, 1994. The principal aim of the meeting was to evaluate the U.S. and other national climate action plans which had been released a few weeks before.

Specifically, these papers concentrate on an overview of the U.S. plan; the perspectives of business, industry, electric utilities, and environmental organizations; mitigation actions in various plans; integrated assessment; an overview of plans from various nations; and the need to amend the convention.

The meeting was sponsored and cosponsored by 35 governmental agencies, environmental groups, industrial organizations, and educational institutions.

Many thanks are due to the distinguished members of the organizing committee who drew up the program and suggested the speakers; to the speakers who so ably presented the information contained in this book; to the staff of C.E.I. for handling the correspondence, the registrations and the mechanics of the meeting; and to those who helped to underwrite the costs.

The book would not have made it to the publisher without the hard work and cooperation of the assistant editors William Wagner and Wendy Petry.

And lastly, I must thank my wife, Ruth, for her patience during my preoccupation with getting this manuscript to the publisher.

James C. White
Ithaca, N.Y.
July 1995

CONTENTS

1

THE INTERNATIONAL NEGOTIATING COMMITTEE PROCESS

Oscar Avalle

Argentine Foreign Service
Mission to the U.N.
1 United Nations Plaza
25th Floor
New York, New York 10017

I want to talk about the global perspective of the INC negotiations process. To understand that, we have to understand the purpose of the United Nations in the field of development. I am quite sure that if I ask about the United Nations, immediately Security Council issues are going to come up, Bosnia and so on. But what is the role the United Nations can play in an issue where all institutions come to mind, specifically institutions which are based here in Washington? For example, World Bank, IMF, and IDP in the case of inter-American development banks.

So why are we getting involved in the United Nations with things like climate change, particularly if we consider that the decisions taken in the context of the General Assembly are not binding? We have to understand that the United Nations plays a very, very essential role in establishing the trends and establishing the international agenda for certain subjects. The issue of the environment is essential, and I would say that if the United Nations did not exist, we would have at this stage neither the concept of sustainable development nor would we have an understanding that it is necessary to link the issue of economic development, not only in developing countries but also in developed countries, because developed countries also have to develop themselves further.

Sustainable development, which is basic, has not been defined precisely. That reminds me of a quote of St. Augustine, which I learned at school. "What is the Holy Ghost?" Well, everybody knows what it is, but nobody can exactly define what it actually means.

So something similarly happens to the concept of sustainable development. Nobody knows exactly what it actually means, but everybody understands what it refers to. In that sense, what we have been doing in Agenda 21 was essential as a landmark on how to proceed to establish a general framework, after which the bureaucrats decide what we actually are going to do.

And that's why I am eager to listen to you, because the real implementation of what we agree in the context of sustainable development will be, without any doubt, in your hands, not in ours. If we are not able to produce real commitments in the real world and real actions, nothing is going to be done.

Evaluating Climate Change Action Plans, edited by White
Plenum Press, New York, 1996

I am not going to repeat the commitments which are in the Climate Change Convention. You know them much better than I do, because a lot of you have been working already from the point of the real world.

We are getting to a crucial point in the process because now we bureaucrats have to stop talking, because the INC is over. In February, the INC will have its last session. And what we have to do in April is working in the real world. The first Conference of the Party is starting. The question is: Where do we stand now? At this stage, in the context of what was actually agreed to in the framework of the Convention, we are actually in quite good shape. Although this might be overstating the role of the INC, I would say that is true.

First, we are working on the methodology to report on the commitments assumed in the Convention. We already have received 16 communications of developed countries regarding how they are implementing their commitments.

This is very important, because it is essential for the COP to be able to work and to know precisely what we are comparing, because, as we all know, when we start comparing apples with bananas, only a fruit salad comes out, and it doesn't really make sense at the end.

What we now have is the ability to compare things. On the other hand, what is happening in developing country situations in this matter? The developing countries are not obliged to report in the first three years after the entering force of the Convention. That is true, but it is also true that the developing countries are getting interested, as I demonstrated in the last INC, and are committed to establish a framework under which to report.

The problem of developing countries in this context is basically a very simple one. Developing countries don't want to have interference in their internal affairs, which could jeopardize their development. I don't think that there is any developing country which actually is interested in destroying the environment. That would be suicidal and absolutely crazy.

The real concern of developing countries is an obligation to keep the standard of living of their population, of their people, as good as possible.

I am not talking about having one Cadillac or two Cadillacs. I am talking about having food or not having food, that's actually the problem. Am I putting things in a little bit too extreme fashion? But that's the political decisions with which a developing country decision maker is confronted .

In the last INC, there was a commitment by developing countries to proceed on the study of reporting mechanisms. But, of course, there is a problem. Developing countries are ready to study and prepare inventories in relation to what had been agreed in the context of the Climate Change Convention.

But we have to be realistic. We have to pay for it and we need cooperation to implement those commitments. I am not saying that we are going to present, as developing countries, a bill saying, "Well, you guys have to pay it," American taxpayers or developed countries taxpayers. The idea is a little bit different. The idea is that we need cooperation of knowledge and technology transfer in the real sense so that we can improve our methodologies and can actually then compare our emissions to yours and your inventories to our inventories.

And G77, which is one of the actors in the negotiations and the representative of the basic consensus group of developing countries, is committed to a report of a workshop which was held among G77 members in Geneva. The main concern on the table is clear but there must be some provisions for financing and technical cooperation.

The theoretical framework is there, and the developing countries are open to cooperation. We have to translate a theoretical commitment into practical implementation of something which at the end should make sense in the real world.

The second pillar is related to the issue of finance, because it goes beyond the simple implementation of commitments of reporting. We have to think about capacity building. We have to think about the implementation of the commitments which a developing country has, which are in Article 4.1. It has to be very clear what we need to do in that context. The issue of the financial mechanism of the Convention is essential.

I am not going to get into the bureaucratic relationships of the global environmental facility as the operating entity of the financial mechanism and so on and so forth. Of course, if somebody wants to ask a question on that later, no problem. But what is important to understand in this matter is that the relationship is being established and is actually quite clear. This is a very big step forward.

Why is it a step forward? Because that means that we are able to operationalize in the multilateral framework, the commitments regarding the implementation of the Convention.

There are still things open for discussion. One of them is the famous issue of joint implementation. What is joint implementation? That's a question which is one of the hardest questions to answer. Why is that so hard to answer? Because from a perspective of some developing countries, when we get into the issue of joint implementation, there is a real concern that joint implementation would be used so that developed countries would not fulfill their commitments in the stabilization of emissions, which are actually in the Convention.

A lot of developing countries are a little bit skeptical, vis-a-vis the issue of joint implementation but that is an issue which we have to negotiate in INC 11.

The other issue, which is still up in the air, is the issue of the additional protocol of the convention, the adequacy of commitments. Are the commitments enough or do we have to add to the whole international system in the environmental field, in climate change in particular? That is one of the issues which is still open, and one which, in particular, has been addressed by the Alliance of Small Islands — an additional protocol to be negotiated in the INC 11.

If we don't have the input from the real world, in other words from you, then the whole exercise might end up in a very nice treaty, only worth the paper on which it is written. I know that is a very well-known quote, but it is a real quote.

If we don't have the support of the real world in this multilateral negotiation exercises, then the whole exercise is absolutely senseless. History will only say that there has been a Climate Change Convention.

2

THE UNITED STATES PERSPECTIVE

Daniel Reifsnyder

United States Department of State
Office of Global Change, Room 4333
2201 C Street, NW
Washington, DC 20520

This is the second time I have appeared here at the Center for Environmental Information Conference. I could recite the conditions under which I appeared before, but it was some years ago, and things were a little different then, which leads me to a few remarks today about the kinds of things we have been through in terms of the climate issue over the past five or six years.

Several chapters have essentially been closed. I think the first chapter really began with the IPCC and its work toward the First Assessment Report in 1990, which in turn led to the Second World Climate Conference and to the U.N. involvement in the issue ultimately, the formation of the Intergovernmental Negotiating Committee, and the second chapter, which was actually the negotiation of the Framework Convention on Climate Change itself. We are coming now to the end of what I would term the third chapter, which has been an effort within the INC since December of 1992, when it met in its succession. The INC has been meeting to prepare both for entry into force of the Convention and for the first meeting of the Conference of the Parties under the Convention.

We will begin chapter four essentially in Berlin on March 28, 1995, because that is when the first meeting of the Conference of the Parties will take place. As I understand it, it will run until April 7, and there will be a ministerial portion at the end.

So that there is a lot of water gone over the dam with respect to this issue over time. We are here because the Center for Environmental Information has organized this conference. And environmental information is really key to the success of the process that we have initiated under this Convention. It is information and sharing information and methodologies, being able to compare apples to apples, and being able to share and work with each other on Climate Action plans. Under the convention, they are referred to as national communications. There may be some differences between the two. But information and sharing of information is key.

One of the things that I would like to actually read for you today is something that I wrote in a speech back in September of 1992, which is about three light years ago in terms of the climate issue. It really struck me that it sets the stage, particularly for your discussions here over the next three days, of National Action Plans and the review process and what all of this means.

Evaluating Climate Change Action Plans, edited by White
Plenum Press, New York, 1996

In this speech, which I gave in Seoul, I said, "In its brief existence, the INC has been no stranger to strife and discord. Some feel it was miraculous that the committee ever agreed on a framework convention at all. With more countries now signatories to the convention, even than took part in the negotiations that led to it, the INC is a microcosm of the world at large, a microcosm now dedicated to solving what could prove to be the most significant environmental challenge facing our planet. Will it succeed? Will we succeed? A great deal depends on our approach."

"The convention establishes a process in which the parties will develop specific national strategies and measures, and share them openly with others. This process will provide an opportunity to learn from the experience of others and to consider approaches and ways of doing things that might not be readily apparent in a particular national context."

"If the parties approach this process positively, seeking to support each other and build trust and confidence, the results will be dynamic. If, on the other hand, they approach this process negatively, sitting in judgment over each other's efforts and allowing suspicion and mutual recrimination to prevail, the results will be few and ineffective. It is not yet clear which road will be chosen."

That is what I said in 1992. I think it is as true today as it was then. We now have, as Oscar indicated, 16 national communications that have been submitted under the convention. The INC Secretariat has mounted a very rigorous effort to synthesize these plans, and there will be an even more detailed and thorough process of evaluating the measures and the information contained in the plans.

But what is going to be very, very important to the convention, and I think something that people perhaps have not yet given enough real consideration to, is the kind of process it will be. If it is a positive process, I think we will succeed very well. If it is negative, and it could well be negative, I think we are going to have, as I indicated before, suspicion and mutual recrimination, and I think the thing could go no place.

But from the U.S. standpoint, where are we now? I realize I was asked to speak about the U.S. perspective today. We indicated at the tenth session of the INC in Geneva, there were essentially five key items that we identified. Some are related to what I would term the future of the Convention. Some are related to the present, and some are related to both the present and the future. And there is, in a sense, a kind of attention, even in the minds of many of the participants, both environmental groups and the private sector, over whether the emphasis and the focus should be on the present or whether it should be on the future and what the balance may be between present and future.

With respect to the future, the United States called for establishing a new aim to guide our efforts in the initial post-2000 period. We also called for the examination of common measures or a menu of common measures that countries might take to help meet whatever new aim can be established. We noted that the private sector and environmental organizations, in fact, have been not very well incorporated into the Convention process thus far. I think the Convention has been very open to NGOs of all varieties, but it has not really made use of the expertise NGOs can provide.

And we noted that there were certain technical panels that were established under the Montreal Protocol, which have had a very positive effect. We called for consideration of how to involve NGOs more integrally into the Convention process in the future.

We called for strengthening the Convention process itself. It is a bit euphemistic for the effort that has been mounted so far and that I consider one of the great successes of the preparatory phase, which has been establishing the methodology and the format for National Action Plans and for the reports that countries have been submitting.

It took about 18 months to do this in a very determined process, but the results of that have been adopted by the INC and are guiding the efforts of all the countries, all the annex one parties, in developing their national communications and so forth. I think this is very, very important, because if you will recall just a few years ago, in the process that led up to the negotiations, there were a great many targets and timetables that different countries adopted. They were very different if you began examining them in detail. Some of them dealt with CO_2 only. Some of them dealt with all greenhouse gases. Some of them involved gas emissions minus sink. Some of them dealt only with emissions. There were various levels of commitments to those kinds of targets and so forth. And in effect what you had was a real hodgepodge.

We have had the same thing when it comes to being able to compare what countries are doing in the climate arena and with respect to reducing emissions of greenhouse gases and enhancing and expanding sinks. The methodological work that has gone on and the effort to begin coming together around some common approaches is a very significant achievement of the parties in this preparatory phase.

The United States, as I mentioned as the fourth item, had called for strengthening the convention process. We believe that a great deal more work needs to be done, particularly at looking at the effects of measures and trying to measure and agree on how you would gauge the effects of various measures and so forth.

And finally, an item that was also included in our statement that is very central to the equation is how we bring developing countries more centrally into the process. Developing countries currently have quite a number of commitments under Article 4.1 of the Convention. They also have a number of commitments under Article 12.1 and 12.2. Those are the requirements to report on actions they are taking to mitigate and adapt to climate change.

So I am not one that subscribes to the theory that developing countries have no commitments under this Convention. I think it is quite the opposite. They have very significant commitments, but it is going to be how we bring them more into a future regime under this Convention as we begin to consider what can be done, what should be done in the post-2000 period.

This is very critical. I think this emerged very clearly from the public consultation that the State Department and the Administration had, here in Washington on the 3rd of August.

There is a very important question with regard to finances and with regard to capacity building. This, of course, has been a central concern of developing countries throughout the process.

We have now restructured the Global Environment Facility. We have replenished the Global Environment Facility, and the Global Environment Facility, I would say, is off and running. It has had two meetings so far of the GEF council, the last one in October, 1994 here in Washington. There will be another meeting in January. There is another meeting in April. I think there are four meetings scheduled for the first year of the new GEF council's operation. The new GEF has been replenished. It's slightly over $2 billion.

I think these funds and the GEF are very significant, and they are very important to the participation of developing countries in the process of coming to grips with global environmental problems in general. I think they are a very important sign and signal of the willingness of OECD countries and industrialized countries to work with developing countries and bring them into a cooperative process in which we are going to be assisting and supporting their efforts to take actions.

But I think, also, what will be even more important in my mind in the long run to the solution of the climate problem is joint implementation. And I realize joint implementation has had a rather checkered existence; I am not sure it even exists yet under the convention. Let me say it exists. We have had some very raucous debates in the context of the INC meetings. I think that is to be expected. I personally am not as distraught about the kinds of discussions that have taken place as some people have been, because I think it is inevitable.

Joint implementation at the moment has been largely theoretical to most people. They can understand some of the concepts, and they have a number of concerns, which frankly in my mind are largely theoretical concerns. I think what we need to do, and what the United States is in fact doing under the U.S. initiative on joint implementation, is beginning to develop some specific projects, some specific factual information on how these projects might take place, whether the benefits they achieve will in fact be new benefits additional to things that would otherwise have occurred, and whether these are truly going to be win-win kinds of opportunities, both for industrialized countries and developing countries.

I think they will be, very definitely. I think it is important that we get this right. This is what our initiative is trying to do. We have had approximately 25 proposals received thus far under the U.S. program and they are now being evaluated and considered for inclusion under our program. But what it is going to take to sell joint implementation, in effect, are some real life examples of solid projects that achieve real benefits for both sides. With that and with better understanding, we are going to be able to clear up some of the theoretical concerns and the theoretical problems that have existed to date.

So I would conclude by saying I am rather optimistic about the future. We are going to have a very intense period coming up, leading up to INC 11. I know we at the State Department are giving a great deal of thought, with the rest of the Administration, to how to make this a public preparatory process, much as we did with respect to the process we went through for INC 10. In effect, we have been doomed by our success. I think we did a very good job publicly preparing for INC 10, and we are now being held to the same high standards. So it is going to be a lot of work, but it will be very worthwhile.

3

SESSION I DISCUSSION WITH OSCAR AVALLE AND DANIEL REIFSNYDER

Dr. Watson: With joint implementation, as you said, some developing countries, are nervous. Is it possibly that it is the type of project that is making them nervous? Many of the projects that have been mentioned to date have been reforestation-type projects, offsets to emissions in the developed world. I see joint implementation as the potential to effectively achieve elements of sustainable development by developed and developing countries working together to have much more efficient, less polluting technologies and industries in the developing countries.

I wonder if you could comment on if it is all types of joint implementation projects or more what I call the classic forest offset, which really doesn't help sustainable development in developing countries very much.

Mr. Avalle: This question addresses one of the main concerns of developing countries regarding the issue of joint implementation. The first time I got involved in a workshop on the issue of joint implementation I heard about a company talking about joint implementation and making the very strong point that it was a reforestry project.

That kind of project as a main aspect of joint implementation would result in developed countries just creating national parks in developing countries so that they would be able to get out of their commitments. If the issue of joint implementation is directed in the direction of sustainable development, and in particular in the development of new technologies, in the areas of energy in particular, the issues would be totally different. That's my feeling.

It is very important that we get into a pilot phase without credits. Why without credits? Because at this stage what we have to see is if this thing actually works, and if it works, how it's going to be working, in which area and in which direction. If developing countries would be able to get into this kind of arrangement on the basis of no crediting in a pilot phase, the whole process might be working better.

That's one of the issues that we will have to discuss in INC 11. I don't know if we would be able to reach an agreement, but what I do know is that if the issue of joint implementations gets too much into the forestry sector, it's going to be very, very, very difficult, if not nearly impossible, to reach an agreement.

Dr. Watson: This question is for Dan Reifsnyder. What effect will the election of a less environmentally concerned Congress have on the working commitment of the U.S. Government to the climate change issue?

Mr. Reifsnyder: We have been aware for some time as the Administration has sought to deal with the budget deficit and to streamline government in general, of the fact that resources are limited for dealing with any number of problems, including this problem. I don't know whether they will become even more limited in the future or not, but we are used to dealing in an environment in which the resources at our disposal are limited, and we have anticipated that they would remain so. It is partly for this reason that it is so important that we involve the private sector in the solution to the problem. Because, first of all, even if resources were not as limited as they are today, public sector resources are going to be wholly insufficient to truly affect this problem in the long term. It really depends on private investment.

What is going to matter very much is how effective a job we can do at persuading the body politic that the climate issue is one that is not likely to go away very soon and one that requires constant effort and attention.

I have been director of this office since 1989 and I have served in two administrations that have approached the issue from rather different perspectives. And yet, in some ways, there are many elements that are very common, that have not changed very much over time. One is that the United States has been very firmly in favor of a very strong process under the Convention for developing action plans, reviewing those plans, and trying to work with other countries to achieve results.

I don't think that is really going to change very much. We have come out of a recession which has dampened interest, certainly dampened the ability of many countries to take actions, particularly actions that involve public expenditures. In our action plan, we have gotten it right so far. We have relied very much on a partnership with the private sector, on voluntary measures, on working together, an approach that isn't likely to change in the near term. So my final answer to the question is I'm not sure things are going to change that dramatically, but we'll see.

Dr. Watson: Who are the 16 countries that have reported? Does the secretariat have all 16 of the reports? And when will a synthesis of these reports be available for people to analyze and look at?

Mr. Avalle: The last ones were Monaco and Ireland. Before that Australia, Austria, Canada, Czech Republic, Armenia, Denmark, Germany, Japan, Netherlands, New Zealand, Norway, Spain, Sweden, Switzerland, U.K., and U.S. Those are the 18 countries which have been presented. I have also information here that Portugal and Iceland are also going to presented very soon.

The secretariat is going to analyze these documents. It is not considered that these documents are going to be negotiated in any way. It's going to be presented basically in the context of INC 11 and the First Conference of the Parties. The information will be available in early 1995.

Dr. Watson: In document AC237 The G77 in China said, "At this stage, the developing countries need not be involved in a binding agreement on the format for communication of information. Do you want to make any comment on that?"

Mr. Avalle: We have to understand first how G77 works and how this presentation was done in the context of INC 10.

G77 works on the basis of consensus. That means basically it's the lowest common denominator on which a group of 130 countries can agree. In that sense, there were a lot of countries which considered that well, we could go a little bit beyond that. What was important was to send a clear message that there is a commitment assumed by the developing countries in Article 4.1 and Article 12.

It is also true that developing countries have to present their information in three years after the entering force of the Convention.

Taking into account these two positions, the G77 considered that it was important to start an exercise. The exercise, of course, was related to the need to have technical knowledge and not to prejudge the outcome of the exercise which will involve 130 countries.

In that sense, this message from the 77 should be understood not as a negative message, but in such a way that we are not prejudging the outcome of a process which starts now, and has to be finalized at the latest, in three years.

Mr. Reifsnyder: I'm thinking back to the previous question. I was asked about the change in the Congress and so forth. One of the things that has been very interesting to me is that in talking about this issue, and particularly the future of this issue, the adequacy of commitments issue and so forth, has been the very strong interest, particularly from the private sector and in Congress as well, in getting developing countries more engaged in the solution to climate change.

That is a key objective for the United States, and meeting that objective is going to take some resources. I say this because as we try to encourage developing countries to begin developing national communications, it's clear that even where there is willingness — and I think there is willingness on the part of many developing countries to engage in this process — there is a lack of both funds and expertise.

And there's a need for not just in the United States, but OECD countries in general, to reach out and assure that the funding for undertaking this is there and to provide the expertise, in particular to provide the experts to work with developing country experts on methodologies, models, data gathering and so forth. I think this is very, very important.

And we've been doing a good bit of that in the U.S. country studies program. We're now working with over 50 countries in this program. It's been highly successful. We've had countries in the INC actually come out and mention the U.S. program by name, and you have to live in the U.N. world to understand how rare it is to have anybody say anything nice about the United States publicly in a multilateral forum. But we've had quite a number of countries acknowledge the U.S. program and acknowledge what a benefit this has been.

I think the IPCC and the continuing work that the IPCC is doing, particularly as it seeks to complete the second assessment report and to involve developing countries, experts, and scientists in the assessment process, is also important. We have to support the IPCC process with funds.

And finally, the GEF. The GEF funding, as I mentioned earlier, is critical to the involvement and the continuing participation of developing countries in this process.

Now, to the extent that the new Congress looks at these kinds of programs and says, "This is just money, you know, being thrown away internationally," we can save that money. We can use it for other things here at home perhaps but, on the other hand, we will miss achieving the key objective that has been cited to us repeatedly by the private sector and by Congress, which is involving the developing countries more centrally in this process.

You can't have it both ways. And I think the GEF funding, the continued vitality of the GEF, the continued vitality of the IPCC and of our efforts to work cooperatively with developing countries in this area is going to be absolutely critical to bringing developing countries into the process.

Dr. Watson: What are your views on the Cairo Plan of Action on population and will it would affect emissions? Do you think that would have or be taken into account as a measure within the COP or the climate convention?

Mr. Avalle: Regarding the Cairo Population Plan of Action, the linkage between it and environment is basically in Chapter 3. When we negotiated that chapter, we had to bear in mind the issues of emissions and climate change, seen from the perspective of sustainable development.

When somebody mentions Cairo, everybody just remembers population, and nobody remembers the other elements which are in the Cairo Plan of Action, which are basically the elements related to sustainable development.

I think its implementation would have a positive impact on the stabilization of emissions, not only in developed countries, but also in developing countries. And I'm not talking here about the stabilization of reduction of populations. I'm talking about the issue of population and development, and in particular population and sustainable development.

If you go through Chapter 3 of the Cairo Plan of Action, you will see that a lot of the language is actually language used and agreed in Rio. I can assure you that if we implement the plan there will be a positive impact on emissions. In which sense? In the sense that we're going to be implementing sustainable development.

Dr. Watson: What specifically is the U.S. currently doing to make the process leading up to INC 11 and COP positive and avoid splits between Annex 1 and non-Annex 1 countries?

Mr. Reifsnyder: We're working very hard in an informal phase. We're probably going to go into a more formal phase beginning in early 1995 when we begin really working up the specific positions for INC 11 and for COP 1.

There's been a great deal of informal dialogue, interchange, and discussion among various participants and not just among the Annex 1 parties, but among Annex 1 parties and other parties. I've had some very interesting discussions with Brazilian counterparts and others, particularly at an IPCC session, on a very important issue that we have not yet addressed, Article 2 of the Convention.

We also had an IPCC plenary session in Nairobi just recently, and I stopped in Saudi Arabia for some discussions there on my home. There's been a great deal of discussion and exchange of ideas among the different players.

I think everyone is desirous, perhaps the Germans most of all, of ensuring that COP 1 is a success. What it will take to make it a success isn't entirely clear at the moment. I think that there are various views. We have only, as I understand, three proposals now on the table, the AOSIS proposal, which calls for a 20-percent reduction in CO_2 emissions by the year 2005, a German commentary on that proposal, as opposed to a specific proposal, and I gather that there is also now a Swiss proposal.

These are certain ideas that have been floated formally. There have been many more ideas that are being discussed informally among governments.

Beyond that at the moment, I don't know what to say. I think the way we prevent rifts is to talk to each other. And the way we develop consensus is to talk to each other. And that has been exactly what's been taking place.

Mr. Avalle: I would like to add something as a non-Annex 1 country. I think that what has been said is absolutely correct in the sense that the contacts are there. The contacts at the informal level, in many cases in this kind of exercises, are much more important than what we are often seeing in the formal settings.

Those kinds of contacts never ended actually after INC 1, they were a kind of continuum and the other advantage is that the actors in most of the negotiations, in most of the processes, are the same. I mean, we all see each other several times a year, and that kind of contact really helps to reach consensus. By identifying the problems, we can really work together to have a common solution without getting into the rifts and conflicts.

The process is there, and I'm very optimistic that we will be able to keep that process.

Dr. Watson: The next question is, "How many countries, Annex 1 countries, are likely to meet the nonbinding goal of stabilization of emissions in the year 2000 at the 1990 levels?"

Mr. Reifsnyder: It is no secret that meeting this goal has been very difficult, is very difficult, for all of the countries involved. There are some who are perhaps having fewer difficulties than others, but by and large, across the board, I think most of the countries involved have found meeting goals to be very difficult.

And it has to do with a number of things. Some people had paid very little thought to exactly how you could meet this aim ahead of time, in terms of the kinds of measures that it would take, in terms of the kind of domestic effort that would need to be mounted. Since the Convention was negotiated everyone has had to sit down and really get into that kind of a process, and it's been tough.

We have had some success in certain ways. For example, economic growth in this country has been much better than expected but better economic growth also leads to essentially higher emissions than expected.

You don't want to wish for lower economic growth so that you can meet the aim but, on the other hand, the positive things that come out of higher economic growth are problems in terms of trying to meet the aim of the Convention.

Fuel and energy prices are low and have remained low which has been a problem. There have been a lot of different factors that countries have had to try to come to grips with.

One of the reasons that we have proposed work on developing a Menu of Common Actions is our sense that countries are going to be very reluctant henceforth to consider new aims under the convention without a much clearer sense of how they might get there and exactly what might be involved.

It is no secret that the aim that is currently in the Convention was very much a political aim but people are now asking much harder questions, much sharper questions, when it comes to discussion and consideration of a new aim.

I don't think it's clear yet that a new aim would be necessarily similar to or identical to the previous kind of aim that you have in the Convention itself. At a workshop in Brazil, just in October, we began to look at the scientific and technical aspects of the objective under the Convention, which is the ultimate stabilization of greenhouse gas concentrations in the atmosphere at a level that would prevent dangerous anthropogenic interference with the climate system.

There is recognition now that we really need to be paying attention down the road to these atmospheric concentrations and the radiated forcing that may result; dealing with emissions is a means, in a sense, to that end. People are beginning to think more broadly about this issue, which I think is good.

Dr. Watson: I think that largely covers another question I just received, and that was, "Given that most OECD countries cannot now demonstrate they can meet the current commitment, shouldn't the discussion at COP 1 be on how to achieve the current commitment, rather than talking about a future aim." And I think in some respects, you have really touched upon that.

While you are answering that question, I think you said that because we have had most of the world experiencing higher economic growth than forecast, that is going to make it more difficult to reach stabilization at 1990 levels in the year 2000. Couldn't one also argue, though, with a strong economy, one can afford environmental protection much more?

Mr. Reifsnyder: Provided one uses the resources of the economy for that end, yes. Indeed. Indeed, one can.

The question that you asked, though, about shouldn't we be focusing on achieving the current aim and not so much on consideration of future aims, let me say that I hope that we have not conveyed in anything we have said in the INC or elsewhere that we're ignoring the process that is taking place now under the Convention. The national plans, the national communications, the synthesis, the review and discussion and the give and take, the interplay, that's all involved and it's very, very important.

We've been committed to a very strong review process throughout the Convention negotiations and implementation and our interest in maintaining that strong review process continues. It's critical to hold people's feet to the fire in terms of the kinds of commitments that they've made.

On the other hand, it's going to be important that we try to do this in a positive, constructive way. If we start getting into a recrimination type of exercise, saying, "Well, I met my aim. How come you didn't meet your aim" sort of thing, I think that we can create a kind of a spiral downward very quickly. What we are trying to work toward is "Where do we go from here?" The aim that we have now takes us to the year 2000 and really not beyond it. Yes, there are commitments in the Convention to adopt policies and measures, and there is no time limit on the commitment to adopt policies and measures, but there is nothing to guide national efforts beyond the year 2000.

That's part of the reason that it's important to pay attention to this, as well as to how successful countries are or can be in meeting the current aim.

It is important, as people have pointed out repeatedly, that we really work toward a global solution to this problem, and we involve more countries in the effort, particularly countries from whom we are going to see very significant increases in emissions in future years.

Dr. Watson: So what you are saying, then, is the U.S. position is that the current aim is clearly inadequate to meet the long-term goals of the Climate Convention, and we clearly have to have a long-term perspective, as well as a short-term perspective. Maybe you can comment on that and carry on with the last question, "What does having success in Berlin really mean in terms of U.S. policy? Is the U.S. prepared to adopt a goal of a 20-percent reduction in the year 2005?"

Mr. Reifsnyder: The United States has always been committed to a comprehensive approach to this problem. The AOSIS protocol proposal that deals only with carbon dioxide is contrary to the position that we have maintained all along. We've looked at all greenhouse gases, sources and sinks. I think that has not changed.

You know, whether countries will agree to take the kind of action that AOSIS is advocating, I think, is going to depend very much on their estimates of whether they can find the measures to achieve that kind of frankly ambitious target. And my sense is, at the moment, that there's been insufficient exploration of exactly what would be required in order to achieve that level of reduction.

I think that if the measures were there, if countries felt that they could easily achieve them, there would be no problem. In effect, the AOSIS protocol, to some extent, is putting the cart before the horse.

Dr. Watson: The last question that you might both want to comment on — and I will touch on part of it in my talk — the first part of the question, and we will have short answers: "Are the consequences of global climate change too nebulous for countries to take action?"

And the second part of the question is, "Since the U.S. and Canada are the highest emitters per capita, does not this mean that the responsibility for reduction through technology improvement is squarely on North America? And haven't we got responsibility to make sure all countries have access to that new technology?"

Mr. Avalle: Regarding the first issue, I think that even if there's no 100 percent sure scientific agreement, most societies in the world — and I'm talking about societies and indirectly, then, governments — are committed to the concept of sustainable development. It has been shown in Rio. It has been shown in the whole process which was the road to Rio and which now is leading the road from Rio.

Politically speaking, it's one of the main issues in the agendas of each country to promote sustainable development.

There is a clear political commitment, not only from the government but from society, to implement sustainable development and the Climate Change Convention.

Regarding the issue of Canada and the U.S., it is true what has been said regarding emissions. At the same time, they are not the only signatories and the only countries which ratified the Convention. We have at this stage 165 signatories, if I remember correctly, and we have approximately 96 or 95 ratifications. So there are 95 countries which are committed to implement the Convention, developed and developing.

And, of course, developed countries have commitments regarding technology and regarding, also, the issues related to the common differentiated responsibilities. But nevertheless, let's be realistic. Only two countries can be blamed for something. It's a common global responsibility. We have to implement this globally, and the commitments regarding scientific technologies have to be in the context of common and differentiated responsibilities, and not just from two countries, but I would say from several others.

Mr. Reifsnyder: Well, I think it's kind of amusing to be asked whether the issue is too nebulous an issue on which to act, because, you know, I think action takes many forms. Is it too nebulous an issue to spend $1.4 billion a year on global change research?

Is it too nebulous an issue in which to negotiate a global convention where we have over 100 signatories. The answer to that is no, because that's what's happened.

Clearly, there are questions that remain, and the IPCC has been working very effectively trying to identify areas where further work is needed. I know our Global Change Research Program is trying to get at the questions that remain about climate change in a scientific sense.

But action is taking place all the time, and there has been an increasing pace of action, particularly in the last five years with the creation of the IPCC, the creation of the INC, the U.N. involvement, the Convention itself, the continuing U.S. expenditures on things like global change research, the country studies program, and the Joint Implementation Initiative.

If the issue is too nebulous to act, there is certainly an awful lot of action going on. Whether it's too nebulous to take further action and how far we need to go, that's what the issue has been about for all of this time. That's what we were trying to come to grips with in the context of the negotiations.

Recognizing that global climate change is a long-term problem and recognizing that there are other competing environmental concerns and constraints, limitations on funding and so forth, what level of action is prudent at any particular time? I think that depends very much on you. It depends very much on the public appreciation of the problem and on the continuing developments in science and so forth.

Do the United States and Canada have a clear responsibility, particularly in the area of technology? Well, we may have a responsibility. I think more importantly we have a tremendous interest. There is a conference, for example, I think many of you are aware of, that's taking place week after next at the White House on environmental technology. I mean, technology is a clear winner.

I think technology offers many, many opportunities. It may be the best bet yet to come up with solutions to this problem of climate change. We are vitally interested in developing new technology and we're vitally interested in disseminating U.S. technologies abroad. The Administration has been working very diligently on this through the Commerce Department and Department of Energy and others. The Clean Car Initiative is an example of the commitment to developing these new technologies.

So whether or not we have a responsibility in that sense, I think we have a clear national interest in developing and disseminating new technologies abroad. I think that's going to continue. I think that there will be bipartisan support for that kind of effort.

4

SCIENTIFIC STATUS REPORT

Robert Watson

White House Office of Science and Technology
OEOB - Room 443
1700 Pennsylvania Avenue, NW
Washington, DC 20500

I'd like this morning to quickly touch upon some of the highlights of the current state of knowledge in climate change. This is not meant to be a comprehensive discussion. It will primarily focus on the issues associated with one working group of IPCC; that is to say, the physical climate system, and only touch very, very lightly, if at all, on the implications and consequences of climate change.

I'll also talk about how IPCC is going to write a synthesis document. One of the challenges for IPCC over the next year is to write three documents; one on what we know about the physical climate system, working group one, asking what do we know about greenhouse gases, what do we know about how they affect radiated forcing, and how effectively climate might change, both at the global and regional level.

Working group two, which I co-chair, is going to look at the impacts, adaptation and mitigation strategies for all ecological and socioeconomic sectors and we're going to look at some of the economics on the sectorial basis. Working group three is going to look at some of the social and macro economic issues.

From that, we're going to write a synthesis report that relates to Article 2 of the convention. The key elements of Article 2 of the convention, very simply stated, are the stabilization of greenhouse gas concentrations at a level in the atmosphere that would prevent dangerous anthropogenic perturbations of the climate system, with special attention to three elements.

These are: one, at a rate where ecological systems can adapt naturally; two, with insured food production; and three, with economic development in a sustainable manner.

We, in IPCC, will not try to define the word "dangerous." It is quite clear to us, as a group of scientists, that the word "dangerous" is purely a political word. We, the scientific community, cannot interpret the word "dangerous." It goes way beyond science and brings in social values, and it is a political dimension.

All we will do in IPCC is try and provide a framework of information such that policy makers can decide whether or not a projected climate change is likely to be dangerous with respect to food, ecosystems or economic growth. And, in fact, what could be dangerous to one country could be viewed as totally not dangerous to another country. Simple things such

Evaluating Climate Change Action Plans, edited by White
Plenum Press, New York, 1996

as sea level rise, of major importance to all small island states, may or may not be important to many much larger countries.

FOOD PRODUCTION

In a country such as the United States, which has already an overabundance of food, a small reduction in food supply may or may not be viewed as dangerous. But in a country, possibly in Africa, that already cannot feed itself, a small reduction in agricultural production could be viewed as extremely dangerous.

Policy makers need to understand a major issue with respect to climate change. And that is: can you afford to wait to see if human activities are affecting climate and then act? Or do you have to act on theoretical predictions that human activities are likely to cause climate change? In other words, how prudent should we be? How should we apply the precautionary principle?

A few points that I believe policy makers should bear in mind are the following: basically, if you stabilize the emissions of greenhouse gases like carbon dioxide or nitrous oxide that have very long residence times in the atmosphere it then takes centuries to stabilize the concentrations. So it's a very long time period between stabilization of emissions and stabilization of concentrations.

Even when you stabilized the concentrations, the climate itself continues to change and equilibrate for some number of years, anywhere from three to seven decades. So even after stabilization of concentrations, you have a lag time because of the exchange of energy and chemicals between the ocean and atmosphere.

Sea level takes even longer to stabilize and equilibrate. Even once you've stabilized a climate, it takes even longer to stabilize sea level because of the immense thermal inertia of the ocean. It's of the order of centuries.

Also, if there is ecological damage, if a species becomes extinct, that cannot be reversed. If you have significant environmental degradation to ecological systems restoration of those systems, on average, is also decades to centuries.

The other factor that has to be taken into account by policy makers is in essence that the source of many of the greenhouse gases, such as carbon dioxide, is from our energy system, power plants, et cetera. Once one's committed to a power system, coal fired power plants for example, one is not going to retire those prematurely without extreme economic cost, and everyone has to take that into account.

So there's at least half a dozen reasons to remember that if you wait to observe whether or not greenhouse gases do or do not affect the environment, and if they do affect the environment and you don't like those effects, you cannot reverse that environmental damage for decades to centuries. There's a number of reasons why there is an inertia to the system. And, therefore, it's very similar to the atmospheric ozone issue. A complete phaseout of CFCs, which destroy ozone, starting effectively in the next couple of years, will not prevent continued damage to the ozone layer through at least another half century.

The climate situation is exactly analogous. If one waits to see climate change and we don't like it — we're never going to be able to zero out carbon dioxide emissions in the way we can zero out CFCs emissions. It's absolutely impossible given it's involved in both the energy and agricultural systems of the world, but the point would be it would take centuries to recover from that environmental damage.

The way we often think about climate change is that human activities affect emissions and emissions affect concentrations. The way IPCC is going to write the synthesis report is different. We're going to start with Article 2 of the convention. We effectively are going to ask ourselves the question: how vulnerable are ecological systems, socioeconomic sectors, such as agriculture, forestry, water, energy systems, how vulnerable is human health, and what are the implications at sea level rise?

We're going to look at the vulnerability of systems. We're going to look at the response function of how agriculture would respond to a changing climate. We're going to see how it would respond to a changing mean climate. Is it sensitive to the rate of change of climate? Or is it more sensitive to a change in extremes of climate?

How are these ecological and socioeconomic sectors vulnerable? Are they vulnerable to change in extremes, changes in variability, changes in mean, or is it the rate of change of climate that's most important?

We must be sure that we look at the local, national and regional level, not just the global level. The global level is clearly the wrong level. It could well be, even with a significant change in climate, that agricultural production around the world could conceivably stay constant. But there would be significant regional change in agricultural production.

Therefore, from a food security standpoint, you have to assume the trade market will remain viable or even improve significantly over what it is today. So it could well be in many cases there is no global change in a system, such as agricultural productivity, but very, very significant regional distortions.

When we look at the sensitivity of these ecological and socioeconomic sectors, we have to look at it as a function of scale from local to global, as a function of different types of changes; is it temperature or is it precipitation that's most vulnerable?

But in addition, we also have to look at non-climatic stresses, non-environmental stresses, such as: what are the implications of population change over the next century, what are the implications of changes in land use patterns?

Clearly, with a growing population, we must have a significant increase in agricultural production just to feed people. One clearly has to have an increase in agricultural production just to have a good standard of food supply, independent of population. So we have, when we look at the stresses on these systems, to look at non-environmental stresses, such as population and other changes in society.

In addition, we have to look at other environmental changes, the impact of changes in atmospheric composition; i.e., CO_2. Does it fertilize ecological systems and potentially offset the effect of climate change? There are changes in acid deposition, sulphate deposition, and changes in ultraviolet radiation due to changes in ozone.

We will evaluate the sensitivity of these systems as a function of scale, as a function of different changes in climate, and in the context of other changes in society.

Once we understand if a system is vulnerable or sensitive, we can ask: is climate likely to change? We will then start along the more normal thinking pattern, and ask: how can human activities affect emissions? I'll come back to this later.

Once you understand how human activities affect emissions, you can then ask: how will emissions affect atmospheric concentrations? I'll also touch upon that.

For example, with carbon dioxide emissions, whether from fossil fuel or from land use changes, we have to go through a complete carbon cycle to understand how the emissions relate to concentrations. That requires an understanding of the terrestrial biosphere, the oceanic biosphere and, probably most important, the physical uptake of CO_2 by the mix layer of the ocean, with the transfer of that CO_2 into the deep ocean.

Once we understand what concentrations are and the relationship between emissions and concentrations, we can understand the relationship between concentrations and radiated forcing or heat trapping.

We are concerned not only with greenhouse gases, but also on aerosols. We've realized over the last three or four years that sulphate and carbonation aerosols play a very major role in the radiated forcing of the atmosphere. We need to understand the relationship not only between greenhouse gases and radiated forcing, we also have to understand it between aerosols and radiated forcing.

From that we then need to be able to understand how to model the implications of radiated forcing on global climate, and more importantly, on regional climate. The real question then is: do we have to run everything in a large general circulation model that has an interactive atmosphere with an interactive ocean for every trace gas scenario we can envisage, which is many? Or is there a way of finding, let's say, a parameterized fit between what we would predict at the global level and what might happen at the regional level?

To keep running every scenario in a general circulation model, coupled to an ocean atmosphere system is extremely expensive and time consuming. It may be the only way to go at the end of the day, and it could be we will have to use these GCMs completely if we can't find a relationship between global changes and more regional patterns of change.

Once we understand a plausible range of temperatures and other climatic changes, we combine it with the sensitivity response functions of the various ecological and socio-economic sectors to produce a plausible range of impacts. So at this stage, we can combine this part of the cycle with our response function and have a plausible range of impacts for all ecological and socioeconomic sectors, if the information is there. But we recognize that in many cases, we do not have good information.

From that, we can do several things. We can ask: what is the cost of these implications, these impacts, if there are actions? In other words, the decision is: let's just allow the climate system to change. We predict impacts on ecological and socioeconomic sectors. What is the cost, both in market terms and non-market terms? We've got to be very careful not to put all of this in market terms only.

We could also ask ourselves the question: what are the potential adaptation strategies to these changes? And what are the costs of these adaptation strategies in market and non-market terms? We can also ask ourselves: can we mitigate climate change by reducing the emissions of greenhouse gases? And if so, what are those strategies, are they energy efficiency or renewable energies? And what are the costs?

This is a general scheme where we can relate human activities through changes in climate implications and plausible range of impact on systems, looking at the various costs so you can evaluate adaptation strategies versus mitigation strategies.

That's the broad framework that we will use at IPCC. Unfortunately, there is a significant dearth of information on some of these response functions, how different systems will react to different changes in climatic parameters. So we will use this framework but recognize we cannot fill it out completely and that there are significant uncertainties.

Having said that, let's find out what is the some of the latest information. We all know that carbon dioxide concentrations have increased over the last millennia. Carbon dioxide was relatively flat until the industrial revolution around the 1700s, but since then, carbon dioxide has increased significantly, about 28 to 29 percent. It averaged around 280 parts per million in the pre-industrial period. It's now around 355 parts per million.

We believe this increase can be completely ascribed to human activities, both land use changes, mostly deforestation, and the combustion of fossil fuels, with a small amount from cement production.

The major contribution today is from combustion of fossil fuels. But in the earlier periods, around the turn of this century, the larger contribution would have been from deforestation.

The modern record of CO_2 since 1958 is a very spiky one in the year-to-year variability. Over the last few years, the growth rate of carbon dioxide seems to have declined. We do not understand that. The question is: is that an anomalous effect? IPCC argued in the recent document that came out that this was not particularly unusual. If you look in previous periods on a yearly basis, there have been other periods where CO_2 growth rates have been actually close to zero. It looks like there might be a recovery; that is to say, that the rate of growth of CO_2 is starting to go back up. But I think it's premature to make a strong point out of that.

So at this moment in time, one could argue that this slow down is not anomalous, but I think we'll need another year or two to really be able to make that statement strongly or even refute that statement.

One can see lots of year-to-year variability in CO_2, and one of the intriguing questions is: has there been a slowdown in the last couple of years? If there has been a slowdown, do we understand it? Some people have argued that you could attribute this slight slowdown to issues of Mount Pinatubo and El Niño, which by changing temperature and precipitation patterns, affect the exchange of CO_2 between the atmosphere and the ocean and the atmosphere and the terrestrial biosphere.

People are doing some very careful work in looking at the isotropic ratio of atmospheric CO_2 to see if they can ascribe this slowdown to the above changes.

With methane, again, pre-industrial, it was around 700 parts per billion. It's now around 1.7 parts per million or 1,700 part per million. Therefore, it's more than doubled since the pre-industrial revolution. But there is no question that in the last few years, the rate of growth of atmospheric methane has slowed down tremendously.

In the early 1980s, it was going up at more than one percent per year globally. Now it's down a few tenths of a percent per year. There is no question that the rate of growth of methane has slowed down.

There are some speculations in the literature why this is. None of them are very convincing to me. It could, of course, be a reduction in the source strength, or it could be an enhancement in the sink, as we're pulling it into the atmosphere less quickly, or it could be we're taking it out of the atmosphere more quickly through enhanced levels of hydroxyl radical. The latter is not at all convincing to me. The tropospheric chemistry system is a highly buffered system. So to expect a large change in OH in the last decade seems quite implausible to me.

Why would the sources have changed? Again, I don't think we know. Some propose it due to the reduction in emissions from the gas lines in Russia, or a change in some of the high northern ecological systems. It has slowed down; we're not sure why.

To summarize all of this – CO_2 has gone up around 28 percent. Methane has more than doubled. Nitrous oxide has gone up about 10 percent since pre-industrial. The fluorocarbons have gone up infinitely, given that they were not in the pre-industrial atmosphere. They are only industrially produced chemicals. However, as we know, these gases are all controlled by the Montreal protocol, or at least CFC 12, 11 and the other CFCs, such as 113, 114 and 115 are. Even the HCFCs, the hydro-chlorofluorocarbons, are also going to be controlled under the Montreal protocol and effectively zeroed out in 2015. Type 22 is growing at the moment, because it's substituting for some of the long-lived CFCs. CF 4 and other fully fluorinated compounds are also growing.

Now, the thing to note about the fully fluorinated compounds is they have lifetimes of thousands of years. In fact, CF 4 has a lifetime of 50,000 years. They're almost all immortal compounds. Once you put these in the atmosphere, they're there almost forever. We've got to be very, very careful about these fully fluorinated gases.

As I said, there's also been a change in the amount of aerosol in the atmosphere. The amount of sulphate aerosol has increased significantly, primarily because of the combustion of fossil fuels. When you combust coal and oil you get sulphur oxides just in the way that you get CO_2. Over the land masses in the northern hemisphere there's been a significant increase in aerosols due to combustion of fossil fuel. And over some of the areas in the tropics, where you large amounts of biomass burning, you also see an increase in aerosols.

Aerosols have two effects. That is, they can both affect the radiation directly and indirectly. The greenhouse gases have increased since pre-industrial time and we understand their optical properties. Therefore, we can calculate quite accurately the change in radiated forcing or heat trapping due to the greenhouse gases.

We believe that the direct greenhouse gas effect of methane, CO_2, and nitrous oxide is around two-and-a-half watts per square meter. So effectively, we have increased the heat trapping by two-and-a-half watts per square meter by introducing gases into the atmosphere.

Stratospheric ozone has decreased because of CFCs and it has probably helped to cool the system, negative forcing maybe by about a tenth to two-tenths of a watts per square meter.

Tropospheric ozone has probably increased. It doesn't matter if tropospheric ozone increases at the surface, but we believe it's increased in the middle and upper troposphere, where it's a very effective greenhouse gas. The trouble is the observations are quite poor and our ability to theoretically model tropospheric ozone is quite poor. Our best estimates are that tropospheric ozone has increased, and it's increased the radiated forcing by about .3 to .6 watts per square meter. That's additive to the long-lived greenhouse gases, such as CO_2.

But now we come to the role of tropospheric aerosols, SO_2 and carbonates. There's a direct effect of aerosols in the atmosphere, they reflect incoming solar radiation. By reflecting that incoming solar radiation, less solar radiation reaches the earth's surface, hence the cooling.

The global estimate is that there could have been a cooling due to the direct effect of tropospheric aerosols anywhere from about .3 watts per square meter and about 1.5 watts per square meter. It's very, very uncertain.

It's uncertain because we don't understand the distribution of the aerosols very well. There's uncertainty in the optical properties of the aerosols and a few other uncertainties. In the end, they are likely to affect the global average in a range somewhere between .3 and 1.5 with a mean value of about 0.9 watts per square meter.

In addition, these aerosols can affect the cloud optical properties. By changing the size distribution of the droplets in the cloud, they affect the albedo of the cloud. By affecting the albedo of the cloud, more incoming solar radiation is reflected.

Here, the effect is incredibly uncertain. Just to do the cloud physics is very, very difficult. And so the general comment by IPCC is: we believe we understand the sign. That is, that it's likely to cool the atmosphere. It could be anywhere from almost negligible, minus .1, or up by as much as 1.5, even 2 watts per square meter.

In other words, our best estimate is that the indirect effect of the aerosols is comparable to the direct effect of aerosols; that is to say, they're additive, and they both tend to offset the direct effect of the greenhouse gases.

The problem is the distributions of aerosols. We have high concentrations over land masses in the northern hemisphere and almost none over the oceans in either the northern or southern hemisphere. They are primarily in the northern hemisphere, not the southern hemisphere.

Therefore, even though aerosols are clearly a significant offset of radiated forcing, we do not understand quantitatively the implications for surface temperature. You can't simply say we've got 2 or 3 watts per square meter direct forcing increase by the greenhouse gases, and potentially half or an equal amount decrease by the aerosols and assume there will be no climate change.

There are major changes in the regional patterns of the direct greenhouse gases, and hence positive forcing and negative forcing, we don't fully understand the impact. We still believe, however, there will be significant radiated forcing. The trouble is we need to have much better theoretical models, GCMs, that couple in the effect of aerosols throughout to quantify the exact implications for surface temperature.

Not only have we observed that greenhouse concentrations have changed significantly in the last hundred years, we know they're likely to continue to change. The key determinant that affects emissions of greenhouse gases is population. Therefore, the implications of the potential agreement in Cairo clearly are simple. If indeed there is a lower population growth than we thought, then there will be lower emissions. There is obviously a direct relationship between population, because of its demand for goods and services, and emissions of greenhouse gases.

Other factors that affect emissions are: gross GNP or GDP; technology development and the diffusion of those technologies into the marketplace; land use practices; energy intensity [the amount of energy used per unit of GDP]; energy prices and energy availability – these last are absolutely critical. I understand that some of the new forecasts for coal prices are significantly lower after the year 2000. It's much more available than we thought at a much cheaper price. That will have a major impact on energy prices, and hence probably demand for coal.

Carbon intensity of energy consumed, trade and labor productivity, these are significant determinants which affect projections of emissions over the next century.

Each of these determinants has uncertainties. There's coupling between some of these determinants. Trying to predict what the projections of greenhouse gases over the next hundred years is likely to be is uncertain. However, there's been a large number of these estimates. The estimates by IPCC are effectively unchanged in the recent 1994 assessment and there's a range of plausible projections of greenhouse gases, carbon dioxide, methane, nitrous oxide and the other gases over the next hundred years.

While there is a central tendency among all of the models to the central projection, you cannot say with any certainty that there is a probable scenario. So with CO_2, there's a plausible range of scenarios, based on key determinants, that goes from basically no growth over the next hundred years to a very significant growth, a factor of about six over the next hundred years. So a very wide plausible range of CO_2.

IPCC's statement, if I remember it correctly, says that this envelope probably encompasses at least the 75th percent of the low and the high case. That is to say you cannot exclude even something higher, and you can't exclude something lower.

Now, these are non-policy interventions. In other words, the low case is a case of low population, low GNP, high diffusion of new technologies, including nuclear, but not in response to a climate change scenario. This is how society may evolve on its own.

From this plausible range of emissions, you can come up with a plausible range of concentrations. Going from about 300 parts per million in 1950, in the low case with a flat emission almost equivalent to the global freeze on emissions, you would still expect growth in CO_2 over the next hundred years. And if there is a large increase in emissions, the concentrations could rise to as much as 900 parts per million. So a plausible range of emissions gives you a plausible range of concentrations.

If you did have an absolutely constant emission at the year 1990, CO_2 would increase from about 350 parts per million today to almost 500 parts per million by 2100, and it would actually rise to something close to 550 parts per million by about 2200.

So even constant emissions would mean the atmospheric CO_2 would increase significantly because of it's long lifetime, and it would basically approach twice pre-industrial levels.

What we also did in IPCC was to say: what if we wanted to stabilize atmospheric CO_2; i.e., Article 2 of the convention? Article 2 doesn't tell us at what level we should stabilize concentrations, so we said: what if we wanted a stabilization at today's level, 350? What if we wanted it at 450, 550, 650 and 750? What would it take to get there?

Several things that are very clear. At any level, whether it's 350, today's level of CO_2, 550, twice pre-industrial CO_2, or 750, almost three times pre-industrial CO_2 and more than double today's, by the year 2100, the projections are that you would have to have lower emissions than are being set.

And if you wanted one of the lower emissions profiles, it's very significantly less. But not only is it less in the year 2100, but after 2100 the emissions would have to drop off very, very significantly. They would all have to be lower than today's level, even for stabilization at three times pre-industrial.

So the message there is: if you want to stabilize anywhere from today's level to 750 parts per million, after about 2100, you have to have very, very low emissions. But even over the next century, you have to have emission levels significantly lower than proposed.

The other way to look at it is there are many, many paths and there is no unique solution to any of these paths. To get to any one of these stabilization levels, there are a multitude of paths. But what we found was there was robustness and what we called an integrated amount of CO_2 that you could put into the atmosphere over the next 200 years. So if you wanted to stabilize at 350 parts per million, today's level, over the next 200 years, you could only put into the atmosphere somewhere between 300 to 500 billion tons of carbon. If you wanted to stabilize at 750, the other extreme range, you could put in between 2,000 and 2,500.

Now, how does this compare to what we're currently thinking we're doing? For stabilization at 350 over the next 100 years, over the next century, you could put in 300 to 400 gigatons.

The central scenario, suggests that over the next 100 years, you would put in 1,500 gigatons. So in other words, stabilization at 350 is significantly less. It's almost one-quarter to one-fifth of the proposed amount. But it's not only that. It's one-half of the lowest IPCC scenario. The lowest IPCC scenario was 770 gigatons over the next hundred years. Stabilization at 350 would require basically half of that. So, in other words, a significant decrease in emissions compared to today.

Stabilization at 550 would be about 800 to 1,000 gigatons, actually somewhere between the two lowest IPCC scenarios. Even stabilization at 750 would allow you to put between 1,200 and 1,400 billion tons of carbon in the atmosphere over the next hundred years.

So the message really very simply is: if you want to stabilize at any of these levels between today's level and even three times pre-industrial, the emissions over the next hundred years have to be on average less than even the lowest of the IPCC scenarios; that is to say, not just stabilization of global emissions, but reductions of global emissions. That obviously is a policy decision of whether one wants to stabilize at any of these levels or something lower or higher.

Now, what are our predictions of climate change? These still come from the 1990 and 1992 assessments. They've been reevaluated for the comprehensive 1990 assessment. But at this moment in time, the general thinking still is that the climate sensitivity, the global delta TG — global temperature change at two times CO_2 at equilibrium — still is one-and-a-half to four-and-a-half degrees Centigrade with a most likely value somewhere around two-and-a-half degrees Centigrade. This is the temperature change when you've double CO_2 and you wait for the system to come to equilibrium. It is a totally hypothetical situation. A climate sensitivity of 2.5.

What would we predict the change in temperature could be in the year 2100? We have taken all of those plausible scenarios of IPCC, all the way from the one that was a global freeze, to a factor of six fold increase in CO_2 over the next hundred years. We combine the plausible range of emissions with the plausible range of climate sensitivity, which is 1.5 to 4.5, and come up with a plausible range of delta temperature globally of 1 degree to 4.5 degrees. The lowest emissions profile, combined with the lowest climate sensitivity, yields a predicted one degree temperature change in the year 2100. If I take the highest emission profile and combine it with the highest climate sensitivity, I come up with a number around 4.5.

This comes out of Tom Wigley's simple model that tries to allow for the aerosol offset, both sulphate and carbonaceous. He's tried to allow for the fact that sulphate emissions will increase, but they will not increase as much or at such a fast rate as CO_2. We've tried to allow what I call a slight aerosol offset.

Some people could argue the aerosol offset is probably larger than in this particular model. In our model, you would get this plausible range of one to four-and-a-half. In Tom's model, if you ignore the aerosol effect, it's not a much different range. It's only about 1 to 5 or 5.2.

One might ask what could a plausible range of temperatures be for the USA? If you look at literally all of the general circulation models, they tend to suggest that the surface area of the USA will warm up at a rate faster than the global average. The number ranges from 1.2 to about 1.7. I use an average value of about 1.2 to 1.3. So to convert my global numbers of 1 to 4.5, I multiplied by about 1.3 to try and find what is a plausible temperature range in the USA.

Again, I've combined plausible climate sensitivities, plausible trace gas scenarios, with a scaling factor from global to U.S. temperatures. If I don't allow for the aerosol effect, I come up with 1.2 to 6.7 degrees Centigrade or obviously up to 10 degrees Fahrenheit. If I do allow for the aerosol effect, it's somewhat suppressed. I come up with a number like 1.2 to 5.8.

But this assumes that the global aerosol offset equals the U.S. aerosol offset, and that is highly unlikely in my opinion, given that most of the aerosols are likely to be over the land masses. If anything, this is almost certain to underestimate the aerosol effect.

The model of Wigley's attempts to take the aerosol offset into account. It's a simple first attempt at doing it and it could be that the offsets are larger than Wigley suggests. The numbers to bear in mind are, if you don't allow for an aerosol offset, 1 to 5 degrees globally, 1-and-a-little-bit to 6 degrees over the U.S.. With an aerosol offset, you suppress it. The key question is how much you suppress it.

In another calculation Wigley tried to calculate the climate change over the next 500 years. Obviously, when stabilized at today's level, there is only a small climate change, a few tenths of a degree Centigrade. This of course is globally averaged.

Why do you see a change from today, if you've got the same concentrations as today? The answer is simple. The climate is equilibrated to today's atmospheric concentrations of CO_2, and we would expect some small change. That's the so-called unrealized warming of past emissions. That change is something like a few tenths of a degree Centigrade.

If we went up as much as 750 parts per million, one predicts something like a 3.5 degree Centigrade rise, quite consistent with those other numbers. As far as sea level, one could potentially get as much as a 20 centimeter to 120 centimeter sea level rise for those particular scenarios.

The interesting thing about sea level is that even after temperatures stabilized by the year 2100 or 2150, sea level will still be going up. The reason is the very, very large thermal inertia in the ocean. It takes a long, long while to realize that sea level rise. Hence, it would take a long, long while if you even stabilized temperature and reduced temperature to get sea level to go back down again.

PREDICTIONS OF PRECIPITATION

All the predictions say you get an increase in precipitation globally. If you've got a warmer surface, you heat up the oceans more, you get more evaporation. So simply, it's got to come down. You would predict changes regionally, but these predictions are quite uncertain.

There is a prediction again that it is relatively certain that you would have a greater frequency of intense rainfall. So even if you had the same rainfall, it might occur in more intense storms rather than being spread out and being gentle. Most of the models predict a mid-continental dry-in. I would have to put this as not a likely scenario, it's got a significant amount of uncertainty.

The combination in some of the models of either the same or less precipitation, but certainly warmer temperatures, means you have more evaporation. Therefore, even if you've got the same precipitation or slightly more precipitation with the higher rates of evaporation, you could get mid-continental dry-in.

But this is where we need the models to be much more sophisticated, to see if there is robustness in these current conclusions.

On the sea level, if I take again my plausible range of scenarios and plausible range of climate sensitivities, one would predict sea level in the year 2100 to be anywhere from 6 to 92 centimeters above today's and roughly half of it, would come from thermal expansion of the ocean. Some of the other would come from melting of the land glaciers. And to be honest, it's very difficult to predict what's going to happen in the Antarctic and the Arctic.

Have we observed climate change? The answer is yes. Over the last hundred years, we've seen an increase of somewhere between 0.3 and 0.6 degrees Centigrade. We found that much of this warming in the northern hemisphere over land is at night, not during the day. This clearly will have different ramifications for ecological and socioeconomic sectors than if it was during the day. But it doesn't mean to say that it's benign.

Thermal expansion of the ocean doesn't care whether it's a daytime temperature increase or a night time temperature increase. From the perspective of thermal expansion of the ocean, it's irrelevant.

ECOLOGICAL SYSTEMS

Do they care about heat stress during the middle of the day? The answer is yes. Therefore, if indeed we're not increasing the maximum temperature during the day, that means we're not putting more stress on the plants. But by having warming temperatures at night and in the winter means that the plants actually can't relax. Having higher temperatures at night may be potentially just as detrimental to the ecological systems as having heat stress during the day. This is another area where we need much more research.

So we do have to look at the implications of a nighttime temperature increase, rather than just a daytime temperature increase.

In the late 1980s and 1990s the temperatures were anomolously high. That stopped effectively with the eruption of Mount Pinatubo. And we actually observed, after the Mount Pinatubo eruption, a global cooling of about .4 degrees Centigrade. The theoretical models predicted about a .4 to .6 decrease in global annual temperature.

So there's actually a reasonable consistency between observations and predictions. Does this mean to say the models are perfect? No. It says that some aspects of the models must be fairly realistic. The short transit response of a change in forcing to a change in surface temperature has to be about right. But it's a very short transit, only a couple of years. It doesn't tell us anything about the exchange of energy between the mixed layer of the ocean and the deep ocean. It does suggest that some elements of the model are quite good, but it doesn't tell you that you should have faith in all facets of these GCMs.

A big deal has been made about an inconsistency between the ground base record and the satellite record, the argument being that the satellite record shows no real increase in temperature over the last 15 years, and the ground base record does. That's a fact. But if you look at the error bars in the two systems, they overlap each other. So statistically speaking, there is no inconsistency.

I think if the record continues for many more years, and it shows no increase in the satellite record, then we're going to have to start to look at this even more seriously. These two records are very important. At the moment, they're not incompatible from a mathematical perspective.

The other point that's been made is: why haven't we seen a large increase in temperature in the last few years, when greenhouse gases are increasing? There are several reasons that are possible. One is that we go into a natural cooling cycle that's being offset by a greenhouse gas warming cycle. Over a 10 or 15 year period, you can never deduce cause and effect, because you don't understand what the system would do in the absence of greenhouse gases.

Another factor, especially in some areas of the northern hemisphere, is that you've had an increase in sulphur dioxide and other aerosols. And that itself has offset the greenhouse effect. That wouldn't be true at all latitudes and all longitudes, however.

Within the USA, over the last hundred years, we've seen about a .4 degree Centigrade rise, roughly consistent with the global average, more at night than day; about a .5 degree Centigrade increase at night, about .3 during the day over the last hundred years.

Since 1950, an even more different pattern. We've seen about three-quarters of a degree increase at night and only about .2 degree Centigrade during the day. It is intriguing that much of it is over the western part of the U.S., not the eastern part. And again, we have to start to look at the theoretical models to see if this makes sense.

The reason why you could get a larger nighttime increase than the daytime increase is simply if you've had a lot of aerosols in the atmosphere, they could effectively be offsetting the greenhouse effect during the day, reflecting incoming solar radiation, but they don't have that same effect at night. So it is plausible.

The general shape of the increase in the last hundred years has been an increase from 1900 to 1940, a decrease over the next 20 or so years, followed by an increase in the last 25 or so years. The overall magnitude is not inconsistent with observations. Clearly, the GCMs don't predict exactly the shape but it's potentially not inconsistent when you take into account the aerosol offset.

At this moment in time, this pattern can certainly not be used to verify that the theoretical models are correct, but we would argue in IPCC that this general pattern is not inconsistent with the models. If you take into account an offset for the aerosol it would generally suggest a climate sensitivity of 2 to 2.5 degrees Centigrade change for 2 times CO_2 equilibrium.

That's really what I would like to talk about. It is clear that one of the emerging issues is what are the implications for ecological systems. There the most likely issue might be the rate of change of temperature and climate and, in particular, rate of change of extreme events.

Through IPCC, we're going to try and analyze very carefully what the evidence is about the natural adaptation of ecological systems. Some of the paleo records would suggest that ecological systems would prefer not to experience rates of temperature change any more than 0.1 degree Centigrade per decade, whereas many of these theoretical calculations are significantly greater than 0.1 degree Centigrade per decade.

We're going to look at the paleo records to see what is the evidence of natural adaptability and migration to changes in temperature and we have to realize that ecosystems do not migrate as a whole. That's actually why Article 2 of the convention is probably not written in a very careful way.

Ecosystems don't adapt, species adapt. As you have a change, even if ecological systems could migrate north with an increasing temperature, the systems would dis-aggregate themselves and then have to re-aggregate. And when they re-aggregate, they will be different ecological systems with different mixtures of species.

And so systems will not naturally adapt basically. I think the question is: will you get a significant change in ecological function and a significant change in ecological structure? That's really the key question that one needs to understand and we have to understand the relationship between structure and function. That is, if you change the species composition, will that ecosystem change its geochemical cycling - the way it cycles water and other ways it affects goods and services?

Human health is becoming an interesting issue, and if there is a change in temperature, or change in climate, rather, there could be an increase in heat mortality. If there's a change in extreme events, and I want to stress this because the models are not at all definitive, this could again affect human life, as well as all structures. But not only that, if there's a change, you could actually change the range of vector-borne diseases, like the aerial extent of malaria. You can also change the areas in which you would have cholera and other diseases.

So the health community starts to look very aggressively at the implication of climate change on human health, all the way from heat and mortality to malaria, cholera and other infectious diseases.

We have to get improved scientific knowledge, clearly. There's a lot of scientific uncertainty. There's uncertainty in the time and magnitude in regional patterns of change. There's uncertainty in response functions of ecological and socioeconomic sectors. But what one has to realize is that all the models predict a change in climate of varying degrees, and the long lifetime of the system, from the atmospheric lifetime of CO_2 to the infrastructure of the turnover capital stock, argues that you cannot wait for cause and effect to be established. Policies must be made now, whatever those policies are. But no action is a policy decision.

So governments, industry, everyone has to work together on policy actions today, given the long time frame of the system.

5

DISCUSSION WITH ROBERT WATSON

Question: A small but influential number of scientists remain skeptical about climate change as a significant concern. How do you respond to the skeptics?

Dr. Watson: Indeed there is uncertainty. There's no question about that, uncertainty both in understanding the physical climate system; in the carbon cycle; in the consequences of systems to climate change.

But I do believe there is a lot of evidence to suggest that climate change is likely to occur. The exact magnitude is not well defined. Many people like to believe that uncertainty only goes in one direction, they say, with overestimated climate change. That may not be the case. We may have underestimated implications of human actions on climate.

And therefore, I believe that it is prudent for society — that's environmentalists working with industry, working with governments — to see how one can buy time, how one can use cost-effective actions of energy efficiency, prudent use of renewal energies, in order to get a better scientific knowledge, in order to move forward. We must remember these very, very long time scales of the system.

Putting our head in the sand and not addressing these issues, I believe, could be a very serious mistake. We only have to look to see what happened on a much simpler system, the ozone system. In the 1980s, many scientists and policy makers said, "There's a lot of uncertainty in ozone." They were absolutely right.

Over the succeeding period of the mid-1980s through literally the last couple of years, every time we turned around, the issue got worse. We understood the Antarctic ozone hole. We found there was also loss in the northern hemisphere in winter. Now we've got definitive evidence that except for the tropics, ozone depletion is absolutely global in all seasons of the year.

So uncertainty goes in two directions. Prudent actions now, I believe, will keep industry and all of us with more options and more flexibility in the future. So a simple answer is: yes, there are uncertainties, but uncertainties can go in two directions.

Question: Why are the aerosols in the southern hemisphere skewed towards the west when wind patterns should skew them more to the east?

Dr. Watson: To be honest, I don't know the answer to that. I understand the wind patterns and the way it would skew the northern hemisphere slightly to the west of the continents. Why it should skew it slightly to the east of the continents in the northern hemisphere, why it skews slightly to the west in the southern hemisphere, I'm not sure. But

I was intrigued that this was as much to the west in the southern hemisphere as it was. So I don't have a good answer for that.

Question: Can you give us some idea about how you're going to approach the difficult area of response functions?

Dr. Watson: Clearly we're going to have to do two things. One, we're analyzing the literature as well as we can at the moment. There's fairly decent literature on agriculture, on a whole range of crops. How do they respond to a change in mean temperature? How do they respond to a change in ultraviolet radiation in some cases? How do they respond to precipitation? We don't have a response function for complicated natural systems, and, unfortunately, we have not put an adequate amount of research effort into understanding these systems. The U.S. Global Change Program is now trying to put more emphasis on looking at the consequences of climate change, rather than just trying to look at the physical climate system.

So part of the answer is: we're looking at the literature as well as we can. We will certainly not overstate what we know. We're going to have to show these response functions with very, very large uncertainty bars. There's no question about that.

We also need to increase the amount of effort in the research agenda in these particular areas. So again, when we come up with these response functions, they will have some significant uncertainties associated with them, especially for the natural ecological systems.

Question: You don't define "dangerous," but will you define "significant"?

Dr. Watson: The answer is no. I could leave it at that glib answer but it's a social judgment of what is significant. What is significant? Half a meter sea level rise in Bangladesh will have a significant displacement of people.Would it be significant probably for Holland? No. They've already got sea walls that can deal with another half meter. Plus with their technical engineering, they'll just build them higher.

There are consequences to wetlands and other things. The issue of significance is a political judgment. It's not a scientific judgment.

So again, I wouldn't want to define "dangerous." I wouldn't want to define "significant." What we will try and explain is the plausible range of impacts to different countries for different sectors; i.e, what might happen to agriculture in the U.S., what might happen to agriculture in parts of Africa. Then the politicians are going to have to decide whether that is significant or dangerous. And when they do it, they won't just look at one sector; they will have to add up all the sectors. For one country, how will it affect my food supply? How will it affect the health of my people? How will it affect sea level and dislocate effectively water supply, et cetera, et cetera?

So I don't believe "significant" is any more a scientific word than "dangerous." We will try and show, if the literature supports it, what the response function is and if there are nonlinearities. Some systems may have a linear response to change; i.e., agricultural production might conceivably go down in a linear fashion. A one degree temperature change could have X percent change in food production. Two degrees might double it.

But there may be some systems with very, very nonlinear functions. If you go beyond a certain temperature change or precipitation change, you may get into a very significant or nonlinear region.

Is the IPCC looking at positive feedbacks? The answer is yes, as much as the literature will allow that. The one that was particularly noted here is: as temperatures increase, what could the effective methane emissions be from permafrost? There's a lot of thought that if we were to increase temperatures, there could be a very significant increase in the emissions of methane from permafrost.

The trouble is the literature is relatively poor. It's an idea that's been around for probably five or six years. I'm not sure the scientific basis is much better today than it was five or six years ago.

Obviously, while some people don't believe IPCC is an unbiased process, I do. And we are trying to therefore look at both the positive and negative feedbacks between climatic change and biogeochemical cycling, trying to look at the effects of temperature on both photosynthesis and respiration, and therefore looking at both uptake and release. And that will be done by working group one.

Question: Some have argued that the recent outbreak of the plague of the India was the result of the intense heat waves in that country, which is evidence of impacts of climate change. Do you agree and is this a good example of potential regional impacts?

Dr. Watson: I honestly don't know enough about the recent plague outbreak. There's no question that the outbreak of the Hanta virus in the southwest of the USA was definitely attributed to a change of climate. The different climatic parameters of temperature and precipitation increased the number of rodents that carried a disease, which was then transmitted to humans.

While I can't say whether this particular outbreak of plague was at all related to any climatic effect, that's the type of issue that we need to look at very carefully, to understand the relationship between climatic parameters and vectors of disease, whether those vectors are fleas, whether those vectors are rats or whatever.

So that is indeed the type of research that's going on, and I think it's something that needs far more attention.

Question: "Will working group 2 evaluate impacts based on three times CO_2, as well as two times CO_2, since this scenario may very well be likely?"

Dr. Watson: The answer is probably no in the following sense. We can look at response functions over a wide range of temperature, precipitation change, and CO_2 in the atmosphere, as much as there's information in the literature.

The problem we found is much of the work in the literature has tended to focus on the two times CO_2 world. So as much as we can analyze the literature to look at the implications of the effect of two times CO_2 or three times or four times, we will look at it. The problem is I'm sure there's a really large literature for us to investigate.

So we will try and look at response functions of systems to ranges of CO_2 and hence potential climatic change, but I think again we're limited by the amount of information in the literature, to be quite honest.

Question: People sometimes assume that ozone depleting compounds have little effect on climate, because their net impact on global radiated forcing may be relatively small. Is this assumption appropriate? Can ozone depleters have significant climate impacts even though little net change in global radiated forcing occurs? Is there any evidence that ozone depleters already affect climate?

Dr. Watson: What we do know is that the ozone in the upper troposphere and lower stratosphere is an important greenhouse gas. We know that CFCs and halons have decreased ozone primarily between about 15 and 24 kilometers. There's no question that that decrease in ozone does change the radiated forcing of the atmosphere, and we believe it will have a cooler impact on the surface.

The only person that's done a careful general circulation model of the atmosphere that tries to relate a change in ozone in the lower stratosphere to a change in radiated forcing, to a change in surface climate, is Jim Hanson at GIS. And he tends to get what you would expect. That is to say, it doesn't have a disproportionately small or large effect.

In other words, if he has an increase of radiated forcing due to CO_2, say a four watts per square meter, and he gets a certain temperature effect, when he gets his ozone effect, which is a decrease in forcing, it almost scales linearly. So what we would have thought would have been the offset due to ozone is just about what Jim gets in his model. But only one model has ever been run. My view is we need other models to be run.

However, having said that, I believe the more important emphasis is not only to understand the implications of stratospheric ozone depletion on surface temperature, but also to understand the implications of the aerosols on surface temperature, both sulphate and carbonaceous.

Question: How many trees does it take to support a human being with the oxygen-carbon dioxide exchange? Or how much green foliage does it support?

Dr. Watson: I haven't the foggiest. I'll be quite honest. I'm sure someone knows the answer, but I don't. So I hate to be glib, but I really don't know the answer, and there's no point in saying five.

Question: What is the evidence that atmospheric CO_2 takes centuries to stabilize? Is this based on observation data or just simulation?

Dr. Watson: It's actually both. We are trying to understand the exchange of air between the atmosphere and the top surface of the ocean – the mix layer. There's a very rapid equilibration between the atmosphere and the mix layer. It's only a few years. The people would argue somewhere between two and five years, maybe three and five years. The real issue is the equilibration between the surface and the deep ocean.

There have been a number of tracer studies, part done with carbon 14, part with CFCs, the fluorocarbons. They're not soluble in water, so they act as a perfect tracer. And there are other tracers.

Based on some observations and based on ocean circulation models, we think that it takes centuries for equilibration between the top mix layer, and the deep ocean. There is some modeling backed up by some a observations of tracers in the oceans.

Question: Regarding the framework for IPCC synthesis statement, given the speculative nature of effects and the linkages between physical and socioeconomic effects, what is your view on decision analysis and integrated assessment models?

Dr. Watson: I'm not sure if the effects are speculative. I think the effects are real. Quantifying them is where it's not speculative. I think there are large uncertainties in the quantification. I believe the integrated assessment models are a useful tool, but they're only as good as the information we put into them. That is, it's only as good as the construct for that model. I mean, it's only as good as we understand the processes, whether they be natural science processes or economic processes.

Clearly, there are uncertainties in the GCM component, as I pointed out. Clearly, there's uncertainty in the plausible range of emissions. There's clearly uncertainty in all of these factors.

One has to be very, very careful. I think the integrated models can be useful in two dimensions. One, I think they can be very important in identifying what I call critical scientific uncertainties, so they can help to guide our research program. But I also believe they can be useful in looking at plausible ranges of climate change and plausible economic costs, whether it be of inaction, adaptation, or mitigation.

And so as long as you propagate the errors through the system and you effectively understand the limitations of the input to these models, I think they can still be a useful tool. Policy makers need to understand what is the most likely case, but also what is the uncertainty associated. These models, used by the best experts who understand the limitations of all of the input parameters can still be useful.

That's why I tried, in a very simple way, to propagate the errors. And as you can see, they're large. I can propagate the plausible range of emission scenarios, which may not even bound the complete range, and can at least come out with plausible ranges of temperature and precipitation.

And there is no question that what you might find is on the bottom end, very little impact and very little economic dislocation. But if the top end of those scenarios is also true, you could end up with a predicted plausible large change in climate with plausibly large changes in ecological and socioeconomic impacts. It's a useful exercise, but you have to understand the uncertainties in the input parameters.

Question: Climate changes versus meteorological conditions. Please discuss the differences of impacts due to meteorological conditions; that is to say, annual crop yields, infestations, et cetera, and those due to climatic changes; sea level and eco system modification.

Dr. Watson: I think this is indeed the right question to ask, and this what I hopefully will start to address in IPCC. And that is: What is the annual or seasonal variation in crop yield, in eco system viability? In other words, natural changes in the climate system, how do they affect society? And how does that compare with long-term changes in climate?

I think that's one of the right questions. There is no question that there are significant changes from year to year in productivity of crops. We have to ask; if you have a long-term climate change, is that significant? And again, it comes back to a political question. Would a five-percent reduction be significant? Would ten or twenty percent?

I think you can turn it around to say that we can show today that our society is extremely vulnerable to variability in climate. When El Niño occurs without the forecasting, crop yields in Brazil and Chile and Peru plummet dramatically, leading to decimated crop yields.

Now we can predict a year in advance when an El Niño onset occurs. That has affected the planting and the crop rotation, the crop selection in Brazil and Chile and Peru, and they've actually made themselves much more immune to natural climate variability.

So we know we're sensitive to climatic change. The key question will be: Are the long-term changes more or less important in the short-term fluctuation? I believe in some cases the long-term changes are going to be quite important. That's what we're trying to evaluate through IPCC.

6

THE UNITED STATES ACTION PLAN

An Overview

Karl Hausker

Office of Policy, Planning, and Evaluation
U.S. Environmental Protection Agency
401 M Street SW
Washington, DC 20460

I will discuss the National Communication submitted by the U.S. under the Framework Convention on Climate Change and will also offer some thoughts on the national communications that other countries have submitted.

So let me first talk about the U.S. communication. As we know, international environmental agreements are built largely on trust, although they are bolstered to some degree by verification measures. The Framework Convention on Climate Change is no exception. It is a good constitution, which can be strengthened and modified as countries see fit. When it was negotiated, a central pillar was the development of national communications or plans. This was agreed to by all nations. The submissions of these communications this fall gives us an opportunity to see how all countries are interpreting the convention. We can see the actions they're taking and the commitments they're making. We can learn what measures work and don't work in different national settings.

Overall, compliance with the convention has been extremely high. At this point, 15 of the 18 parties who were required to submit communications have done so. The only countries who have not submitted are Iceland, Portugal, and the European Union [EU] as a whole, although most EU members have submitted their own individual communications.

Recently, the Secretariat of the Intergovernmental Negotiating Committee [INC] began an initial review of these communications. Their findings will be released at the February 1995 meeting of the INC. More importantly, the parties to the Framework Convention have agreed to in-depth national reviews of each country's plans over the coming year. This process should help build confidence in the convention and make the actions of all nations transparent.

The U.S. took the process of its national communication quite seriously, and we produced an excellent submission. The evidence of this is that the INC Secretariat requested additional information from 14 out of the 15 countries that submitted communications. The U.S. was the only country that was not asked to give more information, because we submitted a very thorough and well-documented communication.

Evaluating Climate Change Action Plans, edited by White
Plenum Press, New York, 1996

Let me turn to the contents of the U.S. National Communication. It consists of the Climate Action Report and two technical supplements, the first being the Inventory of U.S. Greenhouse Gas Emissions and Sinks, 1990-1993; and the other being the Climate Change Action Plan Technical Supplement.

The core of our submission was the Climate Action Report. The meat of this report is the U.S. Climate Change Action Plan, which describes the actions we are taking to return greenhouse gas emissions to 1990 levels by the year 2000. This was actually developed last year, in 1993, ahead of the schedule set by the Convention. In moving so aggressively, this country reestablished its leadership on the climate change issue and sent a signal to the world that we are serious about meeting our commitments under the Convention. Moreover, I think that this was one of the factors that led to such a high compliance rate this year and to the high quality of communications being submitted by the other countries.

In addition to describing the measures we're taking, the Climate Action Report submitted this past fall also takes a comprehensive look at U.S. policy. This includes descriptions of our national circumstances, all of our international activities related to climate change, our financial support to developing countries, our research program, our public education efforts, and, finally, our efforts to anticipate and adapt to climate change. I commend this document to you all, and my office and other agencies within the Administration can provide you copies, if you don't already have one.

I think the communication that we submitted deserves very high marks for the following reasons: It presents a credible set of domestic actions and international activities designed to mitigate greenhouse gas emissions. It contains a commitment to monitor and track our progress toward achieving our goal of returning emission levels back to 1990 levels. It contains a commitment to pursue additional actions in the coming years, if necessary, to meet that goal. And finally, it sets a extremely high standard for thoroughness and documentation for all of these actions.

We reported detailed information on each individual action, including the effects of that action on emissions, the specific objectives and policy instruments involved, how that action interacts with other measures, and intermediate indicators of progress toward our goals. Over time, we're going to monitor the progress of our plan, using our inventory of emissions and sinks as a baseline. In addition, we have stressed transparency and documentation to provide for independent verification of those emissions and sinks.

Let me turn now to the progress we have made toward implementing our action plan. As most of you know, the plan is designed to achieve cost effective emission reductions. When President Clinton established the goal, on Earth Day of 1993, of returning emissions to 1990 levels by the year 2000, he said that not only does he want us to meet that goal, but he wants us to do it in a cost-effective manner which will not create new bureaucracy.

The key government role in the plan, as you know, is to promote new technologies and provide incentives for actions that lead to emission reductions. Because we have taken this win-win approach, the total cost of implementing the plan is far less than what would have been required under a traditional command - and -control approach. A number of our programs and actions are aimed at specific end use categories for energy or other economic activities that affect greenhouse gas emissions. We have just completed the first year of implementation and we have a record of which we can be proud. Green Lights is a good example. Here's a program designed to promote investment in the most advanced lighting technology in the world, which actually encompasses packages of technologies that can cut a company's lighting bill almost in half.

Since we introduced this plan, we've added over 350 new customers for Green Lights to a base that already included over 1,500 businesses and organizations, and our customers are saving millions of dollars per year through this program.

Another example, a program called Waste Wise, is also run by EPA. This program spearheads actions to reduce the waste stream, to promote recycling, and to promote the purchase of recycled products. It started from ground zero a year ago, and we now have over 300 companies signed up to take specific actions ,as well as over 750 individual waste reduction programs now under way.

To cite another example, DOE has a program called Motor Challenge. It didn't exist a year ago. This program is aimed at helping diffuse a new generation of highly efficient electric motors across all sorts of applications throughout the economy. Today, one year later, we have over 800 partners, and we are soliciting sites for 25 showcase technology demonstrations. As you can tell, this program, like the others, has all the signs of becoming very successful.

In addition to these targeted end use programs, the plan contains two programs of a much broader nature. They challenge industry to show leadership on the climate change issue, to seek ways to reduce emissions and, in return, receive public recognition for their efforts.

One of these programs is run by DOE. It is called Climate Challenge. DOE has signed up over 800 electric utilities, representing over 80 percent of the total generating capacity in this country, to make pledges to reduce greenhouse gas emissions. That is an incredible success rate.

A second program, jointly run by EPA and DOE is the Climate Wise program, which similarly reaches out to industries of all types and works with them to develop company plans to reduce greenhouse gas emissions, and to share information among these partners on what the best approaches are. We have now signed up companies that represent over three percent of the total industrial energy use in our country, and these companies have pledged reductions totalling 20 million metric tons of carbon equivalent.

We have also built partnerships with states and municipalities. These are important players in mitigating our nation's greenhouse gas emissions, as is industry. Seven states are currently working on strategies to address climate change, and by the end of next year, nearly half the states in the U.S. will have taken the important step of beginning to track their own greenhouse gas emission inventories. Thirteen states are assessing climate impacts, such as sea level rise, or are demonstrating innovative mitigation projects, such as telecommuting.

On the international front we have launched a program called the U.S. Initiative on Joint Implementation, [JI]. This inter-agency effort, coordinated by the Department of State, EPA and DOE is encouraging greenhouse gas reduction projects in other countries. We hope to conclude the first round solicitation and selection process by January, just a month from now.

One aspect of JI may not have attracted everyone's attention, again indicating the success of this program, is that recently we signed a statement of intent with Costa Rica on JI, signed at the highest levels, by Vice-President Gore and by President Jose Figueres of Costa Rica. This agreement not only paves the way for JI projects between our two countries, but serves as a model for future bilateral agreements.

Finally, the U.S. has outlined several mechanisms for monitoring and tracking the progress of the action plan. We have established a bi-annual review process and are committed to pursuing additional policy measures as necessary.

That is a review of our success which is the optimistic side of events pertaining to the Climate Change Plan. But I would be accused of being a pollyanna if I didn't mention the possibility of fallout from the last election.

The Administration is committed to meeting the pledge it made on Earth Day 1993, but we face a more hostile political atmosphere over the next two years. We are going to be working with the new Congress, and we are going to try to stick to our goals.

There were fears, I think, in some circles almost as early as six months after we first introduced the plan that the plan isn't working and we're not going to meet our target. Well, that was because energy prices were lower than expected in the immediate aftermath of the plan. Now they have firmed up a little bit. Economic growth is somewhat higher than we expected.

We expect to come back to this plan and modify it as needed. There's no way you can set a plan seven years in advance and think that everything is going to go exactly as you predicted. Nevertheless, we are committed next year to conducting the new emissions forecast, seeing how the plans are working, and we will modify the plan as necessary.

One under-appreciated aspect of the plan is that there are pieces that were not scored for emission reductions. We built some insurance into the plan. Three of the programs I mentioned that are now under way, Climate Challenge for electric utilities, Climate Wise for industry, and the State and Local Outreach Programs were assigned no emission reduction tons in our original scoring. We think that this provides a cushion. We are going to assess the degree to which that cushion is going to help us next year.

The early pronouncements that this plan is not going to work were vastly overstated. We have been busy the last year and a half getting our programs in place, getting out there and marketing them, and have not been looking over our shoulder every month and wondering about the latest emission forecast .

Let me turn now to how the U.S. communication stacks up against the communications of other nations. We have done a review of the national communications that have been submitted to date, and would like to share with you some of the insights from that initial review.

In general, we are quite pleased with the thoroughness and the quality of the communications submitted by other countries. Of the 15 communications submitted, they included collectively over 700 specific actions taken to reduce greenhouse gas emissions. Countries have used a wide range of policy mechanisms: everything ranging from economic measures to voluntary programs to regulations to support for our research and development.

For the most part, the plans cover the major sources of all greenhouse gases, not just CO_2. We think that this is extremely positive. In addition, almost all countries have mentioned that they are planning or implementing measures that look at sinks, as well as emissions.

About a third of the countries in their communications addressed adaptation measures. Several countries included energy taxes in their plans, although some of these refer to taxes that were actually in place before the signing of the convention. About two-thirds of the communications indicated support for the Country Studies Program, which is yet another important element of the U.S.'s international activities, a very successful program that builds capacity in developing countries. Six of the communications expressed interest in joint implementation, which is a U.S. led effort.

One country actually developed a backstop in case their current actions are not sufficient to meet the aim. This country is New Zealand. It committed itself to implementing a carbon tax in 1997 if its plan is not on track to achieve its year 2000 target.

Another country, the Netherlands, relies heavily on comprehensive voluntary agreements with entire industry sectors. This is similar to the U.S. approach regarding Climate Challenge and the electric industry, and Climate Wise and the industrial sector.

Despite this good news, there are some areas of concern that we have identified in reviewing the national communications. Procedurally, the INC reporting guidelines were not always followed. Documentation of methods for inventories, projections and measurement of actions were not always provided, as we did in our case. Some parties did not report the impacts of specific actions, only providing an aggregate impact of all their programs. Some countries did not clearly distinguish actions that were in place before the Convention was signed, as opposed to truly incremental actions after the Convention.

More importantly, we are concerned with the baseline calculations by a couple of countries. One country raised their 1990 estimate of emissions to account for higher than normal temperatures in 1990. That's not fair. In another case, a country included emissions due to the import of electricity from other countries, which is not in strict keeping with the reporting guidelines either.

So what can we make of all this, when we consider the national communications? Based on this initial screening, we believe that the reporting guidelines and review procedures should be strengthened before the next national plans are proposed. The U.S. will be pursuing this objective over the coming year.

Overall, we do have reason to be optimistic. The largest Annex 1 parties are responding with national plans or will do so shortly. And these plans, as a whole, are based on the kind of programs the U.S. feels are necessary to meet the current aim of the convention.

That is how things look from the international perspective. Now I would like spend a few minutes discussing the opportunities presented by our own Climate Action Plan.

I note that there are a number of companies and organizations that have come to this conference that are partners with us in various of these actions. The American Gas Association is involved in a program called Natural Gas Star that helps plug methane leaks from the natural gas pipeline system, and we appreciate all their work.

A number of universities are able to sign up for the Green Lights program, and in so doing not only protect the environment but save money. Green Lights is a great example of a broad program that can sign up customers ranging from a multi-billion dollar corporation, to a local school board, to medium-size companies and to universities. Universities, like Carnegie-Mellon, Columbia and University of Virginia are partners, and I would urge anyone from a university which is not a partner to get their university involved in this, with particular emphasis on Cornell. I'm not sure if they have signed up for this program yet, and I would be heartbroken if my alma mater was not ultimately a Green Lights partner.

We have one company who is, I believe, a sponsor, if not a participant, in this conference, Martin Marietta. They are participating not only in Green Lights, but also in Waste Wise and the Climate Wise program, which to me is a pretty good hat trick. And again, I urge companies here: if you are not involved in the Climate Change Action Plan, if your company wants to show any environmental leadership on an issue that is ranked in the "serious risk "category by the EPA Science Advisory Board, you ought to get involved. You can sign up for Green Lights. You can sign up for Waste Wise. You can get involved in Motor Challenge. Any company that has a light bulb, a wastebasket or a motor can get involved in these programs, as well as any of the more targeted programs that lack such broad appeal.

We have available a list of all the programs and contact phone numbers. I would urge everyone in this room to think hard about the organizations they are involved with, including towns or counties or school boards ,and think of what those organizations can contribute. There are opportunities for every type of organization, and every level of government to participate in the Climate Change Action Plan.

I hope each of us will go beyond attending conferences, noodling ideas, crunching a few numbers, and ask: What am I doing to respond to the President's call? What is *my* organization doing to address this issue? Working together, we can meet the President's goal.

7

LUNCHEON DISCUSSION WITH KARL HAUSKER

Question: From a corporate perspective, what are the pros and cons for corporations participating in these voluntary programs? What is the sweetener for them doing it?

Dr. Hausker: I like to think of the programs in two basic categories. One is focused on some end use practice, like more efficient lighting technology, more efficient motors, and plugging natural gas leaks. In many cases, EPA or DOE can provide information that will reduce the transaction cost to a firm to achieve cost-effective emission reductions. These programs are designed to maximize cost-effective opportunities and gives a company an opportunity to save money. A company has an opportunity to focus its energy on finding those opportunities.

In the Green Lights program however, it's very hard for plant managers to get empowered by their senior management to make the investments, even though they may have one to two year paybacks. For a company to focus on this program, and get senior management to empower the plant managers to make the investments, brings tremendous payoffs.

The second category of programs, those that are broadly focused, include Climate Challenge or Climate Wise. Here, we are trying to establish partnerships with companies even though we may not have new technology to recommend. We may not have information that can reduce their costs, but we want to join together with them to help them promote their own ideas on how they can reduce their emissions costs effectively. One of the biggest rewards is, after being spurred to make a self-examination and committing to reduce emissions in cost-effective ways, to then get public recognition as an environmental leader, which we believe does have value to companies. I would urge information on that particular program.

Regarding Climate Wise, we will have a presentation by Pam Herman of my staff today. She has some amazing success stories with participating companies, just as we have success stories in the Green Lights and the Waste Wise programs. When companies examine their internal operations they find that, "Gee, we haven't focused on this before, but we're looking at investments with incredible internal rates of return on lighting or source reduction or recycling."

Question: Could you explain how you propose to track the emissions and keep inventories in coming years?

Dr. Hausker: Yes. Each year we do an inventory of our greenhouse gas emissions. This generally lags roughly a year behind the current year's data but compiling annual inventories gives us a trajectory of where emissions are going over time.

Participant: Who does it?

Dr. Hausker: That is a joint DOE/EPA project. We have a memorandum of understanding between the two agencies to do that on a joint basis. DOE has the most comprehensive information on the energy sector and thus on CO_2 emissions. We at EPA have some of the most comprehensive information on methane and other trace gases. We work very closely together on this to put together an extremely accurate document.

In terms of keeping our eye on the year 2000 target, we have a system that tracks how well our programs are doing and what they are achieving. For instance, it can tell us how many square feet of office space has been signed up for Green Lights, or how many trees are being planted in some of our carbon sequestration options.

So we have a real time sense of program progress. Moreover, we are committed to doing a comprehensive emissions projection every two years for the year 2000 and beyond. We will be looking at economic growth, energy prices,and all the other factors that affect our emissions and then factor in our program success. We'll be doing that next year, just as we originally did in 1993 ,when we produced a baseline forecast and also a forecast accounting for the action plan.

Question: I have two different inventories of emissions, one put out by EIA for DOE and the other by EPA. And they seem to be two different studies. You just said you were working on one. Can you explain this?

Dr. Hausker: The Energy Information Administration of DOE is, a quasi-independent body, and they were given a specific mandate under the 1992 Energy Policy Act to do a greenhouse gas emissions inventory. The rest of the Administration, meanwhile, in complying with the Framework Convention on Climate Change, has developed an inventory which adheres to the internationally agreed upon methods. The methodologies are almost entirely the same, but EIA has gone its separate way on some of the built-in assumptions it used. So there are small differences in the numbers.

Comment: There are a few minor differences between the documents. First,the EIA inventory includes estimates of emissions from bunker fuels [fuels used for international marine and aviation transport]. The EPA report presents those numbers separately, as is requested by the methodologies recommended by the Intergovernmental Negotiating Committee to the FCCC.

Second, the EPA report includes an estimate of carbon sequestered in forest sinks. Again, this estimate was requested by the INC.

Finally, EIA developed a slightly different carbon coefficient for coal. These carbon coefficients were developed too late to be included in the EPA inventory.

8

PERSPECTIVES ON THE U.S. CLIMATE ACTION PLAN

The State Perspective

David C. Rinebolt

Director of Programs
National Association of State Energy Officials
1615 M Street NW, Suite 810
Washington, DC 20036

In August of 1993, State Energy Directors met with Secretary Hazel O'Leary and committed to working in partnership with the Department of Energy to effectuate the Department's policy objectives. We have long considered ourselves the deployment arm of the Department when it comes to new energy technologies, and agreed to work cooperatively with the Department in this area.

The first major opportunity to develop this new partnership was the Green Room, which was established by the Department to develop programs to implement the Administration's *Climate Change Action Plan* (CCAP). The CCAP programs collectively represent a major shift in focus for the Department from a primarily research function to a balanced portfolio of research and deployment activities. The thrust of the CCAP is to eliminate the barriers which prevent commercialization of new energy technologies,paving the way for increased private sector investments in the area. These goals are consistent with state activities under the State Energy Conservation Program (SECP), which uses a small amount of federal funds to catalyze the expenditure of private and public sector capital. Each federal dollar expended under SECP leverages as much as $17 - $25 of non-federal funds.

States also brought to the table valuable experience developed providing restitution to taxpayers who were overcharged for petroleum products during the era of price controls. We developed strong programs which continue to save taxpayers money by reducing energy costs. At least thirty-five state energy offices currently operate commercial and industrial efficiency programs, and thirty-eight state energy offices run residential efficiency programs.

States energy officials participated in the first White House Conference on Climate Change which developed recommendations for programs which could stabilize greenhouse gas emissions at 1990 levels by the year 2000. The focus was to develop a group of primarily voluntary initiatives which would a achieve this goal. The Administration eventually came forward with the forty-four actions which make up the *Climate Change Action Plan*.

NASEO's Board of Directors of the National Association of State Energy Officials (NASEO) and other state energy officials met with the Department of Energy in mid-De-

Evaluating Climate Change Action Plans, edited by White
Plenum Press, New York, 1996

cember to review the initial drafts of the implementation plans for the twenty-two actions within the purview of the Energy Efficiency and Renewable Energy Division. We were, quite frankly, highly critical of most of the proposals because they focussed on process rather than on deployment needs. The NASEO agreed to commit a full-time staffer to the Green Room, which was set-up to coordinate program development and to continue to participate in the plan development process.

We have been extremely impressed with the commitment of the Department to develop effective program plans. Over the past eleven months, the new program plans have evolved to the point where we believe most of them can be successfully implemented.

Six goals have guided the Climate Change Action Plan implementation program development process and the states role in it:

- Implement EPACT;
- Deliver the required emissions reductions;
- Develop milestones and metrics to measure progress;
- Build on existing delivery mechanisms;
- Leverage outside resources to the maximum extent possible; and,
- Follow the original *Climate Change Action Plan* as much as possible.

Following are summaries of the Action Plans which states believe have the most potential to transform the existing marketplace and deliver the emissions reductions assumed during the Climate Plan development process. For additional information on these or any other *Climate Change Action Plans*, please contact DOE's Energy Efficiency and Renewable Energy Clearinghouse at 1-800-DOE-EREC.

The building program is funded at the 8.5 Million dollar level to target energy efficiency in commercial buildings and family housing. The program will provide 6 to 10 grants of $500,000 to $1.5 million to public-private partnerships in communities to elevate the market for energy efficiency and renewable technologies. The community must be a defined geographic area, but can be a neighborhood, a city, a county, a multi-county or urban area, a state, or a multi-state group. The program will also seek to develop 20 to 40 "voluntary" community partnerships which will receive technical assistance but no giant funds. The basic concept is to make energy efficient construction the norm in a community, rather than the exception.

The Community must achieve a 25% average energy savings per building, and must maximize the number of buildings treated. Grants will be balanced among population, community size, and region. Projects must be replicable and each proposal must indicate how it will be replicated.

Projects must be highly leveraged and partners must bring non-federal funding to the table. Technical support will be provided to help in developing proposals and during the project phase. Project recognition and awards are also included.

A six-page summary of the program was recently mailed to potential project partners. The formal solicitation is scheduled to be published in the Federal Register in early December.

Nine million dollars are allocated for the Market Transformation For High Efficiency Building Program which was organized to increase the market penetration of energy efficient building equipment in the residential and commercial markets.

The program sets up a process similar to Consortium for Energy Efficiency's (CEE) super-efficient refrigerator program. DOE has held a series of regional workshops, authorized by EPACT, to identify candidate technologies which would benefit from market transformation. DOE will work with partners and stakeholders to develop implementation plans to overcome market barriers. DOE will also develop consortia to work on project- or technology-specific initiatives.

Because of the timeline for this program, it is likely that Rebuild America communities could serve as ideal markets for technologies targeted under this program. A final list of candidate technologies will be developed during the 1st Quarter of FY '95. DOE will then identify partners and stakeholders and develop preliminary implementation plans for target technologies. Final program announcements are expected during the 2nd Quarter of FY '95. Activities to continue throughout the year.

The Residential Energy Services Program targets new and existing residential structures and provides funding to six existing statewide Home Energy Rating Systems (HERS) which are participating in pilot programs with DOE and HUD. Pilot states are working to improve the linkages between ratings and Energy Efficiency Mortgage (EEM) products being offered by the Federal Housing Administration (FHA), Fannie Mae, and Freddie Mac.

Overall, the program allocated 1 Million dollars, and is focused on the following five objectives:

- Increase the availability of Energy Efficient Financing options;
- Issue National Voluntary Guidelines for HERS (EPACT Section 106) and increase number of statewide programs;
- Develop a builder training and recognition program tied to HERS;
- Provide education and training for realtors and appraisers and alter Multiple Listing Services (MLS) to include energy ratings of homes;
- Standardize forms.

The program is intended to increase the number EEMs, make home energy ratings and energy efficient mortgages a standard part of the real estate market, and also increase the number of statewide HERS programs. States are increasingly seeding collaboratives which include representatives of the shelter industries in their states, with a goal of developing a statewide HERS. States also support utility-based programs. Many view HERS as a way to increase energy efficiency in the residential sector without resorting to fights over codes.

Grants will be made to the six pilot states - Alaska, Arkansas, California, Colorado, Vermont and Virginia - during the 1st Quarter of FY '95. The pilot states will provide technical assistance to states which wish to develop a statewide HERS. Funding for other statewide program operators is anticipated in FY '96.

The voluntary national guidelines required by EPACT should be issued late this fall for comment, and issued as final guidelines in 1995.

Another program is charged to upgrade residential building standards. This program will work with Code officials, architects and the building trades, while focusing on the following activities:

- Provide technical assistance and incentive funding to states to upgrade and implement residential and commercial codes;
- Provide information to promote benefits of building performance to owners and buyers;
- Establish training programs for code officials, builders, and architects;
- Promote model legislation;
- Provide expert testimony in state and local legislative and regulatory bodies and cofunded training of code officials;
- Grants to states with exemplary programs to enhance and transfer those programs.

This should increase the pace of code adoption and updating, increase the quality and quantity of training provided to the shelter industry, increase consumer awareness of the importance of energy performance in structures, and transfer exemplary training/enforcement programs among the states.

The bulk of funding within this program will be provided directly to states in the form of grants. Matching will be required. A solicitation for grant applications based on exemplary programs will be issued in December, 1994. A solicitation for implementation grants will be issued in March, 1995.

The states will also be charged to create a motor challenge program with market transformation strategies. The purpose is to increase the market penetration of energy efficient electric motor-driven systems. The following activities are anticipated under this program:

- Sign up companies for voluntary Motor Challenge Partnerships;
- Develop Showcase Demonstrations, defined by end-users;
- Continue the MotorMaster Information Clearinghouse operated by the Washington State Energy Office;
- Establish a motor systems database using data from the voluntary partnerships and demonstrations;
- Develop public/private consortia to increase the use of high efficiency pumps, fans, blowers and compressors in conjunction with efficient motors.

Voluntary systems guidelines may be developed. States will be asked to solicit volunteers for Motor Challenge Partnerships; details have not been worked out.

The program Accelerate Source Reduction, Pollution Prevention and Recycling: Expand NICE3 Industrial Pollution Prevention Grants Program is an expansion of an existing effort. Funding will increase from the current $3 million to $5.5 million.

The utility initiatives which have state participation involve the promotion of Integrated Resource Planning (IRP) and market mobilization collaboratives to increase deployment of renewable energy technologies. Effective IRP implementation is a necessary precondition for the effective deployment of renewable energy, energy efficiency and natural gas technologies in the utility sector. Work to date at the Department has focused on research to develop an analytical framework for IRP. The states believe the focus of Action Plan #27 should be on providing assistance to regulatory agencies and state energy offices to insure effective IRPs are adopted and implemented by utilities. IRP should not be viewed as an end in itself; rather, it should be used as a tool for making better decisions on the siting and construction of new generation sources and the role of energy efficiency programs in meeting consumer energy demands. We also support the Department's effort to become a more vocal advocate for IRP by intervening in regulatory proceedings where appropriate. This will help shift the focus of IRP from paperwork planning exercises to real outcomes. States have been actively participating in the solar photovoltaics and wind collaboratives which are the backbone of the Action Plan #26 programs. We believe the two existing collaboratives are working fairly well. States also believe the projects identified in the biomass and geothermal initiatives are generally on target. We are concerned, however, that the market mobilization efforts focus directly on technology deployment; we want to see projects in the ground. If there has been a flaw in the existing collaboratives and the renewable energy programs in general, it is the slow pace of the discussions and the focus on process rather than results.

The states are also involved in the Climate Challenge Program which will be discussed later in your proceedings. It's purpose is to develop voluntary agreements with utilities to reduce greenhouse gas emissions. Utilities representing 80 percent of U.S. generating capacity have indicated a willingness to enter into voluntary agreements. Climate Wise will also be presented here in detail. This will solicit voluntary Climate Wise agreements with industrial partners to reduce greenhouse gas emissions. The program will work with a variety of partners to market the program and maximize the number of Climate Wise companies. This should help to develop innovative strategies to reduce greenhouse gas emissions.

A pilot program of 5 state grants of $100,000 per state will be operated to market Climate Wise to potential industrial partners. A series of ten regional workshops will be conducted over the next year. State funding is $2 Million.

CONCLUSION

State energy offices fully support the collaborative Green Room approach being used by the Department of Energy to develop programs to implement EPACT and the *Climate Change Action Plan*. The Department should be congratulated for their efforts thus far. We believe this approach will result in programs which build on existing deployment efforts, avoid reinventing the wheel, and leverage significant additional public and private sector resources. Voluntary programs - the "no regrets" approach to controlling greenhouse gas emissions - can and will work. These programs will result in economic development and job creation in our states by improving the efficiency and productivity of our businesses and reducing energy consumption in homes. A recent analysis of the plan projects the creation of 157,000 jobs by 2000 and 260,000 jobs by 2010.

By linking the creativity of the private sector and state and local governments in developing effective deployment programs with the research capabilities of the Department of Energy, the programs of the *Energy Policy Act* and the *Climate Change Action Plan* can achieve the environmental goals of the Rio Treaty and accelerate the market for new energy efficiency, renewable, and natural gas technologies and services. These programs will elevate the energy marketplace in the United States into the 21st Century far more rapidly than the action of the marketplace alone.

9

THE ENVIRONMENTAL PERSPECTIVE

Daniel Lashof

Natural Resources Defense Council
1350 New York Ave. NW, Suite 300
Washington, DC 20005

I begin by backing up a little bit to talk about the context in which the National Action Plan was developed, because I think that's important when we try to evaluate the plan and its overall adequacy for meeting, not only the near-term commitment that President Clinton announced, to return emissions to 1990 levels by the year 2000, but the ultimate objective of the Climate Treaty itself, which is quite different. That objective is to stabilize the concentrations of greenhouses gases in the atmosphere at a level that prevents dangerous interference with the climate system.

Bob Watson, in a previous paper, discussed the most recent findings of the Intergovernmental Panel on Climate Change, (IPCC), that are relevant to this issue. The key point being that even at the highest level of concentrations that the IPCC considered, which is more than a doubling from today's concentrations and almost a tripling compared with what the natural concentrations of greenhouse gases were in the atmosphere in the pre-industrial period. Even if you were allowed that dramatic, unprecedented increase in greenhouse gases over the next century or two, you would still have to reduce the emissions of carbon dioxide on a global basis to substantially below what they were in 1990. Global emissions must go substantially below 1990 levels.

Given that the developing countries, with currently extremely low per capita emissions — India's emissions are about one-twentieth of what they are in the U.S. on a per capita basis; China is a little bit more than a tenth — we know that emissions from those countries are going to increase at least for some period of time before they can be stabilized and turned around with advanced technologies.

So to reduce global emissions below 1990 levels eventually will require even more dramatic reductions from the industrialized countries.

To make that even more challenging, while there's no scientific consensus through the IPCC on what level of greenhouse gases is dangerous, many scientists who study potential ecological and health impacts, and sea level rise impacts with different levels of greenhouse gases, have concluded that it's essential to stabilize concentrations at well below an equivalent doubling of CO_2 emissions from where they were pre-industrially.

In addition to CO_2,add the effect of methane, which has doubled from its pre-industrial concentration. With this added effect, CO_2 concentrations can't be allowed to increase to anything like 750 parts per million, but instead need to be stabilized at perhaps 450 parts

Evaluating Climate Change Action Plans, edited by White
Plenum Press, New York, 1996

per million, less than 90 parts per million above what they are right now. To do that, the IPCC scenarios show that emissions would have to peak on a global basis within one or two decades at no more than about 20 percent above what they are now, and would then have to decline steadily and ultimately to perhaps a fifth of what they are right now on a global basis.

This is the ultimate challenge that the nations of the world have committed themselves to, in the Treaty, at least in principle. And it's in that context that I want to turn to the U.S. plan.

I agree with the previous speakers that the programs that are in the current Climate Change Action Plan are good programs, that they will move the United States in the right direction, that they'll make a positive difference, both for the economy of the United States and for reducing greenhouse gases, and other environmental pollutants.

However, I believe it's absolutely clear that they're not adequate to achieve the near-term goal, returning U.S. emissions to 1990 levels by the year 2000, and certainly do not put us on a course consistent with the long-term objectives which will require ultimately dramatic reductions in emissions below 1990 levels.

Let me specify why I've reached that conclusion in a little bit more detail. The first thing useful to look at is what's actually happening with carbon dioxide emissions in the United States over the last few years.

Figure 1 shows CO_2 emissions from 1985 through 1994, actual EIA data is used through 1993, but the 1994 data point is based on the extrapolation in the EIA short-term energy outlook.

Emissions, were a little bit lower in 1991 compared to 1990 as a result of the recession, and we've come out of the recession with very robust economic growth and simultaneously very low energy prices. Emissions have grown dramatically, on the order of two to three percent per year, and are projected to grow something like four percent in 1994, compared with 1993.

Comparing emissions now to what the action plan projected, 1994 emissions are very likely to be above the highest level of any year in this decade projected under the Climate Change Action Plan. Under the action plan, it was projected that emissions would peak in 1997 and then decline slightly for CO_2. Emissions in 2000 are higher than 1990 levels

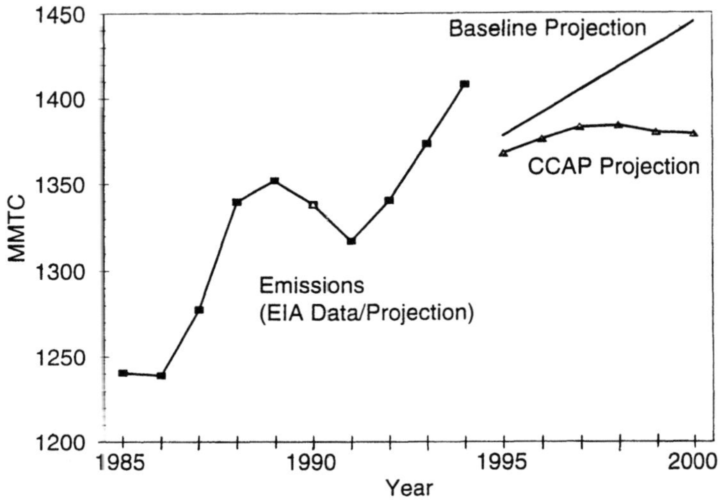

Figure 1. U.S. carbon emmissions 1985-2000. (Sources: CCAP technical supplement; EIA emissions inventory 1985-1990; EIA Data/ACEEE calculation; climate action report).

because the action plan offsets fossil fuel CO_2 emissions with reductions in methane emissions and some increase in CO_2 absorption in trees. I'll come back to that.

This tells us that our challenge is more difficult than we anticipated, even when the action plan was put together a little bit more than a year ago.

I've conducted an analysis to compare where it looks like we're going with where I think we need to go, and believe that we're facing something like a 70 million metric gap between where we're likely to end up under the current Climate Change Action Plan and what would be needed to achieve the President's commitment.

That gap contains two components. The first is that, as I just showed, the baseline that we're working against is substantially higher than what the Administration had anticipated when the action plan was put together. Part of that is more rapid economic growth, part is lower energy prices than were assumed, and part is faulty assumptions in the initial calculations.

And at the same time, it seems to me pretty clear that we are not likely to achieve all of the emission reductions that the action plan has taken credit for.

The first point is that despite the very substantial effort that's been made to put in place a lot of programs very quickly with limited funding, the reality is that we're still behind the schedule that was anticipated in the Action Plan itself, and that the funding is not there at the levels that were anticipated.

Secretary O'Leary testified in May, 1994, as the appropriations process was moving forward, that if the Department's overall appropriations for energy efficiency and renewable energy was cut by more than $150 million, the program would be crippled, emissions would rise dramatically, and the commitment would be virtually impossible to meet. The Congress, in its wisdom, cut the budget request by $180 million.

Now everybody is trying to put the best face on that and say that they can reallocate the money in the most cost-effective ways and maybe the shortfall isn't going to be as bad as you would think initially; maybe it won't be. Conservatively, one would anticipate that the cuts will delay the effectiveness of these programs. I assume a 20 million metric ton shortfall from the rollback of these programs and the savings that accumulate over time will be delayed by one year. I think that's quite conservative, given what's happened with the budget.

The second point here,is labelled "All carbon is not created equal." The plan actually calls for CO_2 emissions to rise from fossil fuel consumption, but offsets that in part by saying that U.S. "sinks" will increase, that we're replanting forests or forests are growing back faster than they're being cut down. And as a result of that, we're going to have 17 million metric tons of additional sinks in the year 2000 compared with 1990. Ten million of that is taken as part of the plan, and seven million is taken as part of the 1990 baseline.

There are fundamental problems with including these measures as part of the plan. Nobody can really tell us the net status of the forests in the U.S. with the kind of precision that is needed to make these estimates. In the last few years, estimates of what the current forests are doing have varied by a factor of two or more.

So there's a big verification problem, qualitatively different from what it takes to track fossil fuel extraction and consumption. Allowing fossil fuel emissions and sinks to be traded off against each other is problematic. But even more importantly, carbon *in* the ground in the form of fossil fuels is fundamentally different from carbon *on* the ground in the form of trees. What this plan actually calls for is digging up carbon, burning it and then putting it in trees. If it's in trees, it can be put back into the atmosphere a lot more easily. Those trees could be cut down in the year 2001, and then all the benefit that we've claimed from enhancing sink capacity disappears. Or they could burn down, perhaps as a result of climate change itself, which is projected to increase the likelihood of forest fires.

I have a fundamental problem counting on trees as an offset for increasing emissions. That doesn't mean that planting trees is a bad idea. It's a good idea. It is part of a comprehensive solution to the global warming problem. The problem is we already needed that to help get us on a trajectory towards lower total emissions. We can't use it up to offset increases in emissions.

In the next section are things that add to the baseline of emissions that we have to offset with the program. These are not shortfalls of the plan itself, but further reflect that the task was harder than was anticipated. Some of these things were beyond the control of the Administration. I think they used reasonable estimates of oil prices at the time, but the world consensus on where oil prices are going is now substantially lower than it was a year and a half ago, when the analysis for the plan was frozen. Based on the current best projections, this adds 16 million metric tons to the baseline.

And I haven't even taken account of the more rapid economic growth than was projected. We don't know that will come out for the rest of the decade, so it's quite likely that it will be a bit higher than what was projected in the plan, but I haven't included that.

The next faulty assumption that was unrealistic at the time the plan was put together is quite significant. The Administration assumed that fuel efficiency of new vehicles would increase substantially to something like 33 miles per gallon by the year 2000 in the absence of any policies and that was incorporated in the baseline itself.

The history over the last ten years is that the combined fuel efficiency of new vehicles, cars plus light trucks, has been absolutely constant since 1985. There are no indications that it will increase. It actually decreased a little bit in the most recent year for which we have data. DOE has consistently projected increases that have not materialized. Not that the technical potential and economic potential to achieve them isn't there, but the market conditions are not there for those kinds of projections to be realistic. If you put in flat fuel efficiency at current levels, that adds about 14 million metric tons to what is needed.

Now, is it hopeless? Is the task too difficult? Should we give up? No, I don't think so at all. I believe that there are plenty of actions that we can take to achieve the near-term commitment and put us on a trajectory towards substantial long-term emission reductions. With the American Council for an Energy Efficient Economy, we've put together our short list of policies to address this gap of 70 million metric tons.

At the top of the list is to increase automobile fuel efficiency standards, a combination of standards and incentives that would raise fuel efficiency levels for cars to 45 miles per gallon by 2005 would, by the year 2000, already lead to a 30 million metric ton per year reduction. But look at the year 2010 number, the big payoff, over 160 million metric tons. That one measure would achieve more reductions in 2010 than the whole plan is estimated to achieve in the year 2000 from other measures.

You might have noticed that in the previous papers about the Climate Action Plan, nobody mentioned transportation. That's because there's essentially nothing on transportation in the plan. There's a measure on tire labeling and something about parking that still hasn't been implemented, but that's it.

The Administration knows that transportation is absent from the plan. We have to give them credit for recognizing that, which is why they have established a federal advisory committee, which I sit on, which is supposed to make recommendations for increasing fuel efficiency, but there have been no proposals to date to actually do it.

It's critical, both for the near term and particularly for the long term. The U.S. projections for 2010, which came out as part of the Climate Action Report submitted formally to the treaty bodies, is the first place where official Administration estimates for U.S. emissions in 2010 are published – the action plan itself had only given the numbers through the year 2000. In it 80 percent of the projected growth through 2010 comes from the transportation sector. So we've got to deal with that sector. We can deal with it. The potential is there.

We have a partnership for a new generation of vehicles, which is aiming at getting tripled fuel economy. And we had a presentation to the Federal Advisory Committee by the engineers working for the big three U.S. companies. The engineers clearly believe this can be done. The problem is, in the current partnership, there's no commitment to actually build these vehicles, and we don't necessarily have the market conditions, either incentives or regulations, to encourage them to market these vehicles. So we've got to do more, we've got to both increase standards incrementally in the near term and create the conditions to really put breakthrough technology on the street in the longer term.

There are other areas where we can do much more. Home Energy Efficiency Programs will be helpful in encouraging home upgrades, but there are real questions as to whether the market penetration that is needed to really achieve the potential in that sector can be achieved with only a rating program and the current nature of the Energy Efficient Mortgage Programs.

At the moment, these last programs are seen by most of the lenders as a nuisance. There's nothing in it for them financially. The programs are inconsistent, and the history has not been good. These mortgages have been available for a long time in this country, but they are almost never used. More needs to be done to ensure their success. I think the things Mr. Rinebolt has talked about are encouraging, but we need a more concerted effort and some incentives for the lenders to actually make it happen.

Efficiency standards are another area recognized in the plan as important, but the plan doesn't go far enough. We have efficiency standards for some appliances. They have been very effective. We can count on them. We need to expand the range of products for which minimum efficiency standards are set, so we can get the junk out of the market and raise the overall efficiency level of equipment.

Reward Efficiency Investments refers to the utility sector. We have had, in a number of states, utilities who have been given incentives through the rate structure to profit from running effective efficiency programs, rather than, as is the case in most states, where utilities actually lose money if the program is good. They make money if they have an efficiency program that doesn't actually save any energy.

The Energy Policy Act said to states that every state should consider adopting similar incentives, but it doesn't actually do anything to ensure that states do adopt such incentives. We believe it's time, if we're serious about climate, to make that a national requirement. Leave it up to the states to determine how to implement it, but make sure that every state implements a policy consistent with the National Association of Regulatory Utility Commissioner's recommendation that energy efficient, least-cost utility plans should be the most profitable plans for utilities.

This is particularly important right now, because the current debate about the restructuring of the utility industry — movements to make it more competitive, which is a good idea if it's done right — has in fact frozen utilities in the headlights all around the country, even in the states that have had the most effective programs, such as California. And at the moment, I would say that even the utility efficiency programs that the plan counts on as part of the baseline are in jeopardy, unless we renew our commitment to effective programs and regulatory incentives for that part of the utility industry that has to be regulated, namely the distribution system.

Finally, recycling, particularly targeting aluminum and plastic, which have the biggest potential for cuts in CO_2; we can do more there than what is in the plan.

Let me conclude by going back to where I started. The challenge that we face from the climate issue is bigger and more important than just returning emissions levels to 1990. We have to deal with the ultimate objective of the Convention, which means reducing emissions dramatically. And while voluntary public-private partnerships are good ideas that we support, they're not adequate. We've got to do more.

The U.S. currently is facing a significant emissions gap in terms of meeting its near-term goal, and unless we take steps, both to address that gap and to address the long-term needs, we're going to be facing a very serious credibility gap, which will undermine our ability to provide leadership in international fora as we move towards the first conference of parties in Berlin.

10

INDUSTRY/BUSINESS PERSPECTIVES

John Hemphill

Business Council for a Sustainable Future
1725 K Street SW
Washington, DC 20460

I'm glad to be here today to talk to you about business reaction to the Climate Action Plan. Time permitting, I also would like to focus on the next steps as well. But first I thought I would introduce you to the Business Council, since we're a fairly new organization, and I suspect many of you have not heard of us before.

The Council is a coalition of businesses comprised of energy efficiency, renewable energy, natural gas and power production companies that share a commitment to enhancing the deployment of cost-effective, efficient and clean energy technologies. We have about 35 members, and we were formed about 2 years ago.

Since our formation, we've been involved in a number of issues, including climate action activities. It is our position that there are real opportunities, business opportunities to provide for a clean environment and make a profit at the same time. That is our message. It's not necessarily an either/or decision. We recognize that there are a number of competing factors that should be and need to be addressed when one is looking at different energy options. Among the factors that should be considered are environmental costs and impacts and means of mitigating these impacts. We believe that the kind of technologies and services that are represented by our membership can and, in fact, do provide solutions to these considerations. We also are finding, not surprisingly, that we are often competing very well in traditional market situations.

We do believe, though, that there are a many reasons why we aren't seeing the degree of market penetration of our technologies that you might otherwise expect to see. Certainly, we recognize that there's a need and a role for governments — federal governments and state governments for that matter — to take actions is make sure that economic and environmental decisions that are in the public interest are taken.

As a recent participant in a meeting of the signatories to the Climate Treaty, the Council prepared a list of policies that we recommended all governments, that are members of the climate treaty, should subscribe to. The Business Council holds that policies, to the extent practical, should be market based, not relying on command and control. We recognize there is a role for regulation, but we should be looking for market approaches, because such approaches are more likely to attract private investment. It's obvious that private capital will be necessary if we're going to successfully address the climate problem.

Second, we think governments need to recognize the interrelation between energy and the environment – that they are not mutually exclusive policy tracks. We further believe

Evaluating Climate Change Action Plans, edited by White
Plenum Press, New York, 1996

that governments must make their domestic energy polices consistent with the Treaty. They shouldn't be subsidizing technologies that are inconsistent with the Climate Treaty.

Third, we believe that governments should also encourage technology transfer and should be encouraging international banking investments that support activities that are consistent with the Climate Treaty.

I think, Dan Lashof did a good job of laying out the short-term concerns of whether or not the United States is going to achieve its year 2000 goal. The Business Council clearly supports every effort to achieve the short-term goal that the US has committed to. But at the same time, I think we need to realize that the purpose of taking action here is not simply to meet a target in the year 2000. Rather, we think it's more important to take a long-term view – to look at what needs to be done to achieve real and meaningful reductions in greenhouse gases. I believe there is reason for optimism because you'll notice that the savings that Dan was showing in every one of his policy recommendations grew over time. These programs become more robust because in almost every case these policy actions do something that changes the nature of capital stock formation. And with the turnover in capital stock, the impacts of these policy actions grow over time.

And that's why we believe that it's important to take action today regardless of the 2000 target, because we need to start having a meaningful change in the formation of capital stock towards cleaner technologies. In this regard, we have put together some recommendations that we have presented to the US delegation to the international negotiations on global warming. They are very preliminary, but we think they are the basis of a good long-term strategy. These policies are aimed at creating a market-pull condition for clean-energy technologies.

We're encouraging the US government to coordinate or work with other developed countries to come up with a market pull program to include commitments to purchase minimum levels of emerging technologies where the economies of these technologies suggest that they have potential to benefit from economies of scale. In fact, we are seeing this happening now in photovoltaics. We're also seeing it with wind, and we believe you'll see it in fuel cells as well.

The Business Council believes that the US government should have their bilateral financial institutions make much more explicit their commitment to sustainable energy projects. We think they should commit to a minimum percentage investment strategy. To date, their actions have been largely limited to rhetoric and not much in the way of action. We think that the World Bank should also make a similar kind of commitment. Both US bilateral and multilateral banks should report their progress in that regard. We further believe that reporting requirement might have a positive effect on their commitment to clean energy.

We also recognize that market barriers are a global problem — we have this problem in the United States, but it's an even greater problem in many other countries. In some countries their financial and regulatory systems are not really well suited to allow for a competitive market to take place. Many countries have very low, subsidized energy prices and/or tariff barriers that distort market signal and capital flows. And we think that the US government needs to do more, both directly and through their participation in lending institutions to try to remove some of these obstacles. That's a brief summary of the Business Council's position on the role of government in addressing the climate issue.

Before concluding my remarks, let me say that we also have taken a position on Joint Implementation, a concept where one country funds a greenhouse gas reduction activity in another country – presumably a developed country would fund some project in a developing country that reduces greenhouse gas emissions. The Council sees JI as an intriguing, potentially economically appealing strategy for reducing world-wide emissions of greenhouse gases and we support it on that basis. But if you start digging into the details of implementation, we see it as a major problem to measure the benefit or to monitor it. We see sorting through these issues as a long-term process and feel that JI certainly should not be viewed as a substitute for taking cost-effective, near-term actions domestically, especially since we see lots of very economical activities here in the United States.

DISCUSSION

...the company saves money through not losing as much gas and has more gas to market.

Mr. Hemphill: I can just add a little bit to that as well. Others have given you excellent reasons why voluntary programs can work, and I think they do work up to a point. They do raise the visibility within the management structure of industries of the benefits of clean-energy programs, and industry does get some credit and recognition for doing the right thing – things that are economic anyway but maybe not economical enough. But the voluntary program may be just enough to push over the decision in favor of taking action. That's a positive result.

Probably another implicit stick, in terms of why companies are participating in these programs, is that they prefer to participate in a voluntary program than to have some command and control type of initiative imposed on them. So there's probably a little bit of that as well.

Now, I think we will have to wait and see whether the voluntary approach is really going to be adequate in and of itself without some additional actions. But I believe it can work, and I think it has and will work so long as the programs are well founded.

Time will tell how effective these voluntary programs will turn out to be.

...long-term rates. How do we fix this situation?"

Mr. Hemphill: Well, that is a fundamental problem of financing power generation. It still exists today. Although, I think we're beginning to see some fundamental changes in how power plants are developed. I think the restructuring of the utility industry is a mixed bag, but I believe one of the benefits will be the increased opportunities for smaller scale clean power plants to be built. I don't think we are going to as many large-scale coal-fired power plants built in the future, and certainly we're not going to be seeing nuclear power plants developed here in the United States. We'll see them developed in other countries to the extent that they're subsidized, although we are hopeful that that won't happen. Obviously, Bob and I might not agree on this point.

I'm not so sure, also, that I would agree with Bob that electrification based on coal-fired base-load power is a greenhouse gas policy. But I do think there are opportunities for using clean energy technologies, especially if there are mechanisms for decision makers to really look at the costs – eliminate subsidies and address the external costs of carbon-based fuels versus non-carbon-based fuels.

Now, for such decisions to be made requires a decision maker to care about such issues and take actions along these lines. I think we're going to see a form of a utility regulation for a considerable length of time, even under restructuring that will allow for such considerations.

So I think there are opportunities, but there also are real challenges. And I think one has to look at mechanisms for internalizing external costs. Some of those mechanisms will be politically very difficult. We had a round of BTU/carbon tax negotiations that resulted in a 4.3 cents per gallon gasoline tax.

I was surprised, Dan, that you didn't give some credit to this tax for improving auto fuel economy. Although I'm not sure that such a modest tax will have much of an impact, I think that we need to see better price signals or otherwise we may miss out on some opportunities to further reduce emissions. But I do not think all is lost or all is hopeless. I think the shift in general will be to smaller scale cleaner power.

Mr. Hay: For Bob Beck – oh, yes, sir?

Mr. Petty: Just a comment. There is something in the 1990 Clean Air Act amendments in the hazard rain section that specifically encourages some incentives for renewable in that bill.

Mr. Hemphill: That's right. There's a cent-and-a-half production tax credit – no, that's Energy Policy Act and I meant to mention this. However at the same time there is not a lot of new incremental generation being built right now because there is an abundance of existing capacity. But where there is activity, renewables are doing quite well, because they can be very competitive. Specifically, wind energy is getting winning a number of competitive bids.

...biggest beneficiaries from efficiency fixes.

Unknown Speaker: There are a whole host of things. You go to rural states, which you might pick as a free rider, yet requirements to clean up cattle manure and eliminate water pollutants may be just the motivation you need to solve a whole lot of methane emissions problems.

I mean, the types of activities that we can undertake are as varied as the emissions sources themselves. And I think we will have a decent hook in every state on some mix of activities that will happen just voluntarily that will eventually get you to the emissions reductions you want to see.

Mr. Hemphlll: You are also going to have some synergy with respect to the regulation of other emissions. Requirements to meet state implementation plans to reduce emissions of ozone or NOX often times results in reductions in CO_2 and other greenhouse gases as well. I am thinking in particular the OTC petition now before EPA to adopt California low-emission vehicle programs is a good example of such a program. If the petition is adopted and implemented, it would reduce greenhouse gas emissions, as well as ozone.

11

BUSINESS\INDUSTRY PERSPECTIVES

Robert Beck

Environmental Affairs
Edison Electric Institute
701 Pennsylvania Ave. NW
Washington, DC 20004

I noticed on the agenda that tomorrow you are going to be listening to Dirk Forrister. Dirk is employed by the federal government. Yet, he's going to talk about the electric utilities. And I'm employed by the electric utility industry, and I'm supposed to tell you about the federal government. Somehow, I think we got this thing reversed. So I'll do as little damage to Dirk's remarks tomorrow as I can, but, after all, I feel much more at home talking about what we do, as opposed to trying to figure out what it is that they have done and will continue to do.

The interesting thing about the 1994 elections was it reminded me about many people who have been working with environmental issues here in this town for too long. It reminded me of the last time the Republicans successfully took control of the United States Senate. When they did that, the incoming chairman of the Environment and Public Works Committee at that time was a gentleman named Bob Stafford, a senator from Vermont. I don't know how many of you know Senator Stafford or had the opportunity to work with him, but he used to have an extremely dry sense of humor, which I always enjoyed.

He used to tell this story about opportunities and flexibility. It was a story about a young fellow who was hitchhiking his way through the beautiful country in the State of Vermont, and he came to a fork in the road. The sign said, "To the state capital, this way," and the another sign had another arrow that said, "To the state capital, this way." And he wasn't real sure which way to go, but he looked up and he saw an old-timer rocking back and forth on the porch and said, "Hey, old-timer." He said, "I'm trying to get to Montpelier. Make any difference which way I go?" And the old-timer looked at him and said, "Not to me it don't."

That's what flexibility is. You're given choices, and sometimes it doesn't matter how you get there; the idea is you gotta get there. So what I'd like to do today is try to bring some focus on how the electric utility industry is going to "get there."

You know, I'm a fisheries biologist by first degree, and scientists have a way of always talking about uncertainty. But in the industry I work for, when I talk about uncertainty, people always say, "Ah, geez. There they go. You know, they're delaying again. Anything it takes. Slow this thing down. We don't know enough. We can't do this. We can't do that."

I'd rather talk for a minute about certainty, especially when it comes to this issue that's called climate change. What do we know a lot about? Well, we know a lot about

Evaluating Climate Change Action Plans, edited by White
Plenum Press, New York, 1996

emissions and we spend a lot of time talking about emissions. And the U.S.National Action Plan talks about emissions. As a matter of fact, it's probably more of an emissions control plan than it is a climate action plan. It's an action plan, but it's an action plan to reduce emissions.

The reason that it's misnamed somewhat is because while we know how to reduce emissions, we really don't know much about climate. We don't necessarily know what those emissions reductions, or even emissions increases, would do to the climate. The average temperature for 365 days a year, from the polar cap in the Arctic to the Antarctic and in between is about 57 degrees Fahrenheit. That's the average temperature.

What good is that? None. It's nice to know but what does it tell us about climate? If it goes to 60 degrees, is that good or bad? If it goes to 54 degrees, is that good or bad?

It's extremes that tell us where we're going to inhabit. We didn't live in Phoenix until we invented air conditioning. We really probably didn't live in Florida, New Orleans, Houston, or a lot of different places until the mosquitos were brought under control.

We need to focus a little bit more if we're really serious about this issue, on what we know about the climate itself. And that's a very tough nut to crack. I don't mean to be facetious. It is a very difficult job. We do know more about emissions. We do know more about what it costs to do certain things. We don't necessarily know as much about the benefits of those things. Remember, we are trying to focus here on the climate.

The position of the electric utility industry in general is that anything that deals with emissions control and climate change ought to be voluntary. It's got to be comprehensive. By comprehensive, I think we're saying that actions must involve all of the greenhouse gases. It has to be global in nature. After all, we're talking about a global climate change or global warming, and it has to be flexible.

In 1973, we projected that oil prices by this time in our life would be about $70 a barrel. We were very excited about oil shale on the west slope of the Rockies. There were a lot of things that we thought were going to happen. Oil this morning, I think, was in the neighborhood of about $18 a barrel. In 1973, oil was about $16 a barrel. Over a 20-year period, I think we can see that projections ofttimes are in error. We need to use projections and models, but we have to understand that they're tools and we need to be a lot more flexible.

What have we done? We have tried to strike a partnership with the federal government. The partnership is called the Climate Challenge, and it got its kickoff last Earth Day. We had a signing ceremony on the Mall, and that got us started.

It was the 24th anniversary of Earth Day, and I have to admit, that on the first Earth Day back in 1970, I was a freshman at Thomas More College in northern Kentucky, and I marched on Earth Day. I'm proud to say that I marched again on the 24th, down to the Mall to the signing ceremony.

And I noticed that it was a big industry event. It was an industrial environmental event. There weren't the marches; there wasn't the big media coverage of thousands of people trying to march to clean up the environment. We have come a long way. That particular day told me of a basic change, a change that my industry and other industries like ours has tried to adopt; we in industry would like to develop more partnerships to actually get things done instead of always opposing getting things done.

We've actually developed a partnership with the government and some of the environmental organizations to promote electric and natural gas vehicles. That's a difference. I think the difference is apparent in the Climate Challenge.

The Climate Challenge is an effort to put industry money, our money, our rate payers' money and our stockholders' money, up front and into the program. We did very little lobbying of the Appropriations Committee this past budget cycle in the federal budget to get monies in the federal government's budget for these kinds of programs. We're going to put our money in there. Whatever the federal government puts in, that's great. We're all for that, but we don't think that we necessarily need their money to get these things going.

We've got an investment pool, which we're going to capitalize in the neighborhood of about $50 million in the first subscription, and we'll probably have two later subscriptions. What is that pool going to do? It's going to invest in electro-technologies, some of the things that were talked about earlier.

What kind of electro-technologies? Well, you all drive cars and ride in taxis, and the cars have paint on them. The paint has to be dried. Most of the time that paint is several coats thick. The drying mechanism used is generally a gas fired dryer. But in a gas-fired dryer, it takes about two hours to dry the paint, and while you dry it a lot of volatile organics get driven off. Anybody have a microwave? You put the car in a microwave dryer, which is an electric dryer, and you can dry that car in six seconds. And when you do it, you trap all the volatiles right there, too. So you get two bangs for the buck.

Now, is it more expensive to do it that way? Probably, but you have to balance six seconds versus two hours, how many cars, fuel, all those kinds of things. Hey, I'm a fisheries biologist. I'm not an economist. I can't give you an answer to that.

But those are the kinds of tradeoffs that we're looking at and we're trying to get some of these electro-technologies capitalized and into the marketplace. Another example. Anybody dine at some of the finer restaurants in Washington? I do all the time. One is called Wendy's. It's not exactly prime rib, but they get it from the same animal. Wendy's has gone all electric. Now, what does that mean? It means they're going to fry things electrically; they're not going to use fossil fuels to do so. Does that mean that the price of a double is going to go up? Well, maybe. But it also means that there will be fewer emissions from all of the thousands of Wendy's that Dave Thomas now has out there in operation. McDonald's has the same thing, an electric clamshell fryer.

Another example is reprocessing foundry sand through electrolysis. What do you do? You take contaminated foundry sand, and clean it up, and again you do two things for the price of one. You remove a potential hazardous waste, the used up foundry sand, and you replace it with nice clean foundry sand that you can recycle over and over again. You still have some waste to get rid of, but far, far less than what you originally had.

These are the kinds of options, the kinds of initiatives that make up part of the Climate Challenge ,the kinds of things you will see us promote. I have had the opportunity to go to China in the last couple of weeks. One out of every five people in the world is a Chinese, one out of every eight is an Indian — there are lots of people in those two countries. They are developing countries. They are not like we are. They do not have the amenities and the things that we have. They would like to, and they are going to develop their power needs and their transmission and distribution needs.

They'll probably start doing it, at least initially, with coal. But there is no reason they have to burn coal in a relatively inefficient stoker boiler when we can work with them through foreign investment, through the World Bank, the Asian Development Bank, the regional development banks, and get state-of-the-art power plants built over there with fluidized bed technology, etc. This is considerably more efficient, considerably cleaner, yet still burns their own domestic coal, and allows them to grow into a higher standard of living and potentially become a developed country instead of a developing country.

Now, a lot of people call what I just talked about joint implementation. We prefer to talk about it in terms of foreign investment. It is a couple of countries working together to try to get something accomplished that raises the standard of living in one country and, to some degree, pays for jobs and keeps the standard of living high in the other country.

And in the meantime, too, if you haven't reduced emissions, you have certainly avoided greater increases. Again, it's a tradeoff. Is it perfect? No. Have we ever claimed that it was? Have we ever bought into the target that you have to stabilize emissions in the year

2000 at the 1990 levels? No. Does anybody know what that dangerous level of emissions is, what the dangerous level of concentration is that is the aim of the Rio convention? No.

The IPCC has said over and over again that they can't answer that. They're working very hard to do so. But we're looking at a very fuzzy target out there. And so I think in the meantime, the kinds of things that my industry would like to see occur is not nothing at all. We need to do things, but we need to do things that are, as I said, voluntary, comprehensive, flexible, cost-effective and global in nature.

12

SESSION II DISCUSSION WITH ROBERT BECK, JOHN HEMPHILL, DANIEL LASHOF, AND DAVID RINEBOLT

Mr. Hay: We have had, from my point of view, a remarkably peaceful panel in the sense that there seems to be a general endorsement of a voluntary approach, the voluntary approach that's been taken in the U.S. plan. There are definitely differences of degree there, concerns expressed that it's not enough, and a variety of things have been listed as additional steps that could be taken: increased efficiency, an auto fuel economy emphasis, stationary sources, efficiency standards improvements, recycling, more commitment to purchasing new technologies, focus on new technologies, that bilateral and multi-lateral institutions need to commit to more specific targets for renewables and clean energy, more foreign investment, and so forth.

I would like to very much encourage anyone from the audience, who has their own perspective on the U.S. Climate Action Plan, that would like to stand up and speak to it. Maybe you have a somewhat more aggressive and critical view.

In the meantime, I have a question for Dan Lashof, "Has the Administration seen your estimates of the emissions reduction shortfall, and do they agree with these estimates?"

Mr. Lashof: Yes. The Administration has seen them. No, they probably don't agree with them quantitatively, but I think that there is at least internal acknowledgement that there is a shortfall. I understand that the Administration's internal estimates suggest the shortfall is about 25 million metric tons, which is substantially less than what I think it is realistic.

But the Administration has not accepted, for example, the notion that forest sink enhancement shouldn't be counted, and that's a portion of the discrepancy. But if you can piece out the discrepancies between the numbers, I think they're in the same ballpark.

The Administration's official response is that the Plan is not going to be left on a shelf to rot. It's going to be revised. They anticipate, it says, a bi-annual review. So the next major revision would be about a year from now. And they are reluctant to, I think, publicly put a number on the shortfall until they're ready to say what they want to do about it, which presumably is going to happen next year.

What I am calling for is not necessarily to accelerate that schedule in terms of an overall revision. Let's get this part of the plan working, but we do need to add things to it. It's essential for the United States to have credibility as we go to the first Conference of the Parties, for us to put forward a down payment on closing the gap, some additional measures that move us in the right direction, even if we don't yet have the full set of measures needed to solve the problem.

Mr. Hay: For Dave Rinebolt, "With many state energy offices greatly reduced or eliminated — New York, one of the more successful state energy offices, is slated for elimination — and the decline of PVE dollars, how are state energy offices supposed to facilitate Climate Action Plan activities? Perhaps less emphasis on voluntary-type programs and more emphasis on competitive solicitations that would include administrative support dollars for state energy offices might be the way to go. Do you agree?"

Mr. Rinebolt: Well, of course I'd like to see more competitive monies available to state energy offices. That's why they hire me. Practically speaking, however, in this budget climate, I don't see that as a great possibility.

I recently went from answering to 19 Republican governors and 31 Democratic governors to answering to 30 Republican governors and 20 Democratic governors. But we have very successful energy offices that operate in states with governorships in both parties.

And the reason that we have successful state energy offices is because they have been able to go to their respective state legislatures and make a very positive case for their abilities to help business and industry and homeowners within those states do things that make sense in terms of energy use.

I should remark, most of the climate plans are competitive. Most of the solicitations are competitive. There are none of these plans that provide 50 state fundings, like the traditional grant programs. We never saw that as a possibility.

We have also been aware of the elimination of PVE money for a long time. In fact, we had surveyed state energy offices in 1989 and found about 2,300 employees. We recently did a quick survey a couple months ago and found the same state energy offices had about 1,800 employees. That's obviously a significant decline. You can track that decline with PVE monies. You can also track it with attempts, which are constant and always ongoing at the state level, to cut the number of state employees.

But I would also contend that we have learned how to be more efficient about what we do with state energy offices. We are a lot more creative in programs and better at spinning them off than we used to be. Once we can convince industrial groups and groups of manufacturers in a particular area that there are approaches that they can use to save energy, we generally find the consulting community takes those over.

Fundamentally, what we're talking about in energy efficiency and renewables is good business decisions that help people make money, and all we are there to do is catalyze them. I'm cautiously optimistic. Maybe that's just because I started my professional career in Washington in 1980, and it's been uphill ever since. We are going to be able to show at the state level that we can do a lot of things and help businesses do things.

It doesn't cost a whole lot of money in my mind to convince somebody to make a capital expenditure that has a payback in two years or less, we're trying to elevate the market for efficiency and renewables in this country. We are trying to make the things that all of us have been talking about as cost effective a normal part of the transaction. We're trying to make a home energy rating a normal part of a housing sale.

If we do that, then we will capture a lot more savings than are anticipated in the plans. If we do that, we will have a lot of reasons for the business communities in our states to want to see us continue. That's how we're going to continue to exist. We can see the PVE monies go away and survive, as long as we continue to deliver valuable services to our customers, who, after all, are our constituents.

Mr. Hay: Another question for Dan Lashof, "How will adopting pay-at-the-pump auto insurance reduce greenhouse gas emissions."

Mr. Lashof: Pay-at-the-pump auto insurance is basically an idea of moving something that's currently a fixed cost of driving to a variable cost. You buy auto insurance, and it's basically independent of how many miles you drive. Your insurance company probably asks you your mileage, and most people make something up so the insurance company doesn't pay much attention to it, because it's not reliable. And so you pay whatever the experience is in this area, $1,000 a year, for auto insurance. That becomes a fixed cost, and you don't see that when you drive.

The idea of pay-at-the-pump is to collect those same premiums, or a portion of those same premiums, on a variable basis. It's more fair, because you pay in proportion to your exposure to accident risk, and it gives you an appropriate incentive to choose the most efficient alternative, if there is one available, and to string trips together, minimize your driving, whatever. The idea is that it will reduce vehicle miles of travel and reduce greenhouse gas emissions for that reason, without raising anybody's actual overall cost.

Mr. Hay: What are the benefits to companies of participating in the voluntary Climate Change Action Plan programs? Does anyone want to comment what the incentives are, whether they're adequate incentives for companies to want to participate?

Mr. Rinebolt: I think the advantage is that you save money in your financial bottom line. We're going to show you how to make a cost-effective investment or give you the technical information necessary to make the investment,and save money on your energy bill.

I heard an example at a recent conference. There were about 13 industrial plants in this company, and they figured out a standard technique that they could use. In one building, the program had a capital cost of, say, $100,000, and it saved about $500,000 a year in energy savings. And they wondered whether it was worth it to do this, since it was going to save them only half a million bucks on $100,000 of capital, then they figured out they had 12 plants and they could do the same thing in all of them. The numbers add up over time.

Why is it the large industrial customers want to change the way the utilities operate? Because they want cheap electricity, because they want lower costs of production. If we can show them how to lower their costs of production through other techniques, I think that there's an advantage for them.

Mr. Hay: The Natural Gas Star Program, which I'm somewhat familiar with, is a voluntary program between the gas industry and EPA to reduce emissions of methane from natural gas systems. The program slightly changes the manager's time horizon and his management activity because it's a very public program, working hand in glove with EPA, with some pats on the back for doing a good job and possibly some crediting.

So a manager who might have to pay an additional first cost to reduce the methane emissions or might have to look at a payback period that's a little bit beyond what his company would normally look at, now has support. And still, you're looking at initiatives that fundamentally are cost effective, the company saves money through not losing as much gas and has more gas to market.

Mr. Hemphill: I can add a little bit to that. I think that, with these voluntary programs, there's some good explanations of why they can work, and I think they do work up to a point. Within the management structure of industries, they get some credit and recognition for doing the right things, things that would be economic anyway. But this results in getting some positive PR. Those are all positives. Companies are participating in these programs because they prefer to participate in a voluntary program than to have some command and control type of initiative imposed on them.

We will have to wait and see whether the voluntary approach is really going to be adequate in and of itself without some additional actions. But it can work, and I think it will work.

Mr. Beck: I think from our perspective the companies that we serve do all the things I just mentioned. Wal-Mart, Wendy's, McDonald's, companies like that, that can manage from one location all of their heating and cooling and other needs with computerization. They can really save some big bucks over the course of a year when you have that many stores around the country.

Number one, from the utility companies' standpoint, first and foremost, we get to keep the customer. Right now, we're in that mode. We're out there competing for different customers. We're not monopolies anymore. So we have to serve the customer's needs, and if they include trying to reduce their bill through efficiency programs and demand side management programs and other things, then that's what we're going to do. That's first and foremost: keep the customer.

Then secondarily, we like being in a positive stance in terms of environmental protection, whether it's climate change or recycling copper wire or old refrigerators or whatever it is. that's a positive environmental good neighbor. We prefer being a positive good neighbor to making the front page as being the dirty old polluters.

Mr. Hay: For Dave Rinebolt, "You did not mention any promising transportation sector programs. The lack of programs is troubling, because transportation emissions will continue to increase, and states have a critical role in transportation polices. Any comment?"

Mr. Rinebolt: Absolutely. There are a variety of innovative programs at the state level to reduce vehicle mile traveled. They're just not referenced in the plans. They are things we were doing anyway as a part of strategies at the state level to comply with Clean Air Act requirements. We've got motivations other than the Climate Plan to deal with transportation issues.

But, you know, at least from what I understand of the politics of the development of the Climate Change Action Plan, transportation was considered too hot an issue to touch. There are politics written into all of this. We're trying to look at the solutions that don't have any political baggage with them and let the advocates who have other motives fight at the national level. At the state level, for clean air reasons, we're going to try to reduce vehicle miles traveled and do things in telecommuting and in the transportation sector. I hope by the time this conference rolls around next year to be able to tell you more succinctly what we're doing in that area.

Mr. Hay: This is maybe a John Hemphill question. "Conventional energy sources, fossil fuels, electrical generation, are or have been subsidized and are financed over long terms of ten to thirty years, based on their depreciation life. Renewable energy is expected to pay back compared to these fuels but does not have the access to long-term rates. How do we fix this situation?"

Mr. Hemphill: Well, that is a fundamental problem of financing of power generation. It still exists today. The restructuring of the utility industry is a mixed bag, but I think it will result in opportunities for smaller scale power plants to be built. I don't think we are going to see the large one gigawatt, coal-fired power plants, and certainly we're not going to be seeing nuclear power plants developed here in the United States. We'll see them developed in other countries to the extent that they're subsidized, and we are hopeful that won't happen.

I'm not so sure, also, that I would agree that electrification based on coal-fired base load is a greenhouse gas policy. There are opportunities for using clean energy technologies, especially if you are really looking at the costs, eliminating subsidies and addressing the external costs of carbon-based fuels versus non-carbon-based fuels.

Now, that does require a policy position being taken on the part of whoever is regulating utilities. And I think we're going to see a form of a utility regulation for a considerable length of time, even under restructuring. It will be a restructured type of regulation.

There are opportunities, but those are real challenges. One has to look at mechanisms for internalizing external costs and some of these are politically very difficult.

But I was surprised that the NRDC didn't give some reduction credit for improvement in auto fuel economy. Actually, I'm facetious. I think that we need to see better price signals or otherwise some policies that will cause proper decisions to be made. But I do think it's not all loss or not all hopeless. I think the shift is toward smaller scale power.

Mr. Petty: Just a comment. There is something in the 1990 Clean Air Act amendments in the Acid Rain section that specifically encourages some incentives for renewables.

Mr. Hemphill: There's a cent-and-a-half reduction in tax credit but it is in Energy Policy Act. There's not a lot of new incremental generation happening right now because of the adequate amount of existing capacity. But you will see that there is a lot of activity in the renewable area, because they can be very competitive, especially wind energy which is getting awards in a number of competitive bids.

Mr. Hay: For Bob Beck, "Given your hearty endorsement of microwave drying technology, will EEI support microwave clothes drying in the recent rulemaking?"

Mr. Beck: In reality, that is out of my league, and I really can't give you an answer. I will get someone in customer services to answer that question.

Mr. Lashof: Well, I don't know EI's position on microwave clothes drying, but I do know their position on water heater standards, where they are vociferously opposing one of the only measures in the Climate Change Action Plan where we were certain of the results. This efficiency standard for electric water heating would require heat pump-based water heating, where you actually get the benefit of electricity in those applications where electric water heating is going on.

Their reason is very simple. When you have that standard, which is extremely cost effective for the customer, it makes the competitive position between gas water heating and electric water heating less favorable for the electrics, because it raises the first cost to be something more similar to what a gas water heater is,but it lowers operating costs.

So, you know, I just want to caution that while we work with EI on a number of issues, we're partners with them in promoting efficient electric end-use technologies, electric technologies as a panacea are often a disguise for marketing electricity, not for reducing greenhouse gases. People need to be aware of that.

Mr. Hay: We've noticed the electro-technology — I think the trademark is ecowatts — activity as well. And from a strictly scientific and objective point of view, we are going through sector by sector, looking at the electro-technologies and seeing which ones truly are energy efficient from a full cycle perspective and truly do protect the environment.

The industrial sector is the only sector where we have pretty well finished the work. And there, we found that the number was somewhat below one percent of the electro-technologies.

Moving on, for Lashof, "There are rumors that the Clean Car initiative may be abolished next year. Do you think there is any truth to this, and what is the likelihood that it will happen?"

Mr. Lashof: I'm not familiar with that rumor, but it's certainly potentially on the chopping block, if you look at statements that at least some of the new incoming leadership in Congress are making, which is basically to oppose anything that smacks of industrial policy. And however you repackage it, it's hard to think of the Clean Car initiative as something other than industrial policy.

I think it would be very short-sighted and unfortunate if the Congress moves in that direction, both in terms of that program and the kinds of programs that we have been talking about here. I don't think this is should be an ideological issue. It's important to remember that the Climate Treaty was negotiated by the Bush Administration, signed by President Bush and ratified by a Democratic Congress at that time, with President Bush the person who submitted it to the Congress for ratification.

These programs are designed to promote partnerships with industry. It would be a very short-sighted, extremely radical ideological view to indiscriminately chop them out. But, quite frankly, that could happen.

Mr. Hay: Without strong national level incentives, minimum performance requirements for energy efficiency, et cetera, and regulations, how do you ensure that some states will not act as free-riders, sitting back and taking few actions to reduce energy consumption and letting a few states absorb the entire burden of greenhouse gas reduction?"

Mr. Rinebolt: What is cost effective in terms of an energy efficiency investment or renewable energy investment, or non-greenhouse gas emissions investment? It varies across the board. Every state is in a different situation. And I think that no matter what state you talk about, there are some activities that we can undertake within that state that make good economic and business sense, that can help in the reductions we need.

I frankly do not think, in this era of rhetoric about unfunded mandates, that there's a whole lot of will to give states individual targets for reductions and ask individual states to meet them. I think that there may well be a few states that do a free ride, but I'll tell you that energy-producing states are often the biggest beneficiaries from efficiency fixes.

There are a whole host of things. You go to rural states, which you might pick as a free rider, yet requirements to clean up cattle manure and eliminate water pollutants may be just the motivation you need to solve a whole lot of methane emissions problems.

I mean, the types of activities that we can undertake are as varied as the emissions sources themselves. We will have a decent hook in every state on some mix of activities that will happen just voluntarily and will eventually get you to the emissions reductions you want to see.

Mr. Hemphill: You are also going to have some synergies with respect to the regulation of other emissions. Requirements to meet state implementation plans and reducing your emissions, ozone or nox, oftentimes results in reductions in CO_2 and greenhouse gas, as well. I am thinking in particular of the OTC petition now before EPA to adopt California low emission vehicle programs. If that is adopted and implemented, it would reduce greenhouse gas emissions, as well as ozone.

Mr. Hay: About 50 percent of the projected increases in emissions from developing countries over the next century are the result of population growth. If their population were stabilized at current levels, what would global emissions be in 2100? Do you happen to know that number?

Mr. Lashof: Emissions can be thought of as a product of three factors. They are all important. We should address all of them. Population times GNP per capita — that's the one we want to increase — times emissions per unit of GNP. If we can stabilize population and reduce the emissions per unit of GNP, then we can continue to have healthy growth of GNP per capita and still be sustainable in terms of our economic development.

Mr. Hay: Right. In fact, there was a related question, "If developing countries were as efficient as the U.S. in using energy per unit of GDP, how many tons of carbon would be saved in 2010?" Do you have that?

Mr. Lashof: That's a trick question. Developing countries, many have very high energy use per GNP ratios, and they have opportunities to become much more efficient as they develop. But it won't happen without their GNP growing at the same time. And, in fact, the more efficient they become, the more it will boost their rate of GNP growth. I don't think there's an easy answer to that question.

It's certainly something that needs to be addressed, and we have been very active in encouraging the World Bank and U.S. multi-lateral and bilateral assistance to focus on helping developing countries become more efficient and reduce their energy to GNP ratios with still somewhat limited success, although I think we're beginning to see some movement in the right direction at the Bank.

Mr. Hay: Do you see any way to avoid ultimate federal command and control legislation , given the need to dramatically reduce emissions, et cetera.

Mr. Lashof: Yes. I think some mix of mandatory programs is essential. That doesn't mean that the voluntary programs don't continue to have a role to play. But clearly a mix that includes things like mandatory increases in fuel efficiency standards for cars is essential.

As an alternative you could also have an overall cap on national emissions and trading within that cap. And I don't know whether you call that command and control or a market mechanism or what. But you need something more than voluntary partnership programs to solve the problem.

Mr. Hay: "With the new Congress coming in that will most likely approve less funding for Climate Change Action Plan-type activities, do you feel that industry-led programs, such as the Super Efficient Refrigerator Program, should be the norm, as opposed to the exception?"

Mr. Lashof: I think that the role of those programs is going to be even more important than it was before. I am not sure it is the exception in the current plan. Again, I'm worried about the status. I mean, from what we have been hearing recently, utilities motivations to participate in those kinds of market transformation programs has eroded in the last year, in the face of uncertainty about the structure of the industry over the future years.

It's critical that we try, regardless of how the industry is structured as a whole, to take steps immediately to ensure that utilities will continue to have an incentive to invest in cost-effective programs like the Super Efficient Refrigerator Program. And the proposals to do that— there's one pending in the State of Washington from Washington Water Power, which would transfer the cost recovery for their efficiency programs to a wires charge on the distribution system and then insulate it from the debate on restructuring more generally, which I think is a real good idea.

Mr. Hay: I would like to point out for those of you who haven't seen the natural gas vehicle outside, please take a look. This is my personal car that I drive to work every day, and I just ran it through the car wash for you. Each of senior staff at AGA have one of these now, and we fuel them on the second floor of our high rise building. It's a neat little car, so I hope you take a look at it.

Mr. Beck: How much did it cost you?

Mr. Hay: This is a conversion, a Ford Taurus that we had done here in Virginia, and it's outrageous. I think it was something like $4,000 for the conversion.

Mr. Womeldorff: I think we're converting for between $2,000 and $3,000.

Mr. Hay: Well, that's a much more reasonable number for today's market. I think we paid a bit more than that, but I could be wrong. But obviously, that's one of the reasons why achieving scale economies is so important to us and why we keep working with the Clean Air Act and the Ozone Transport Commission and all of this to try to get some scale economies.

Mr. Hay: We are going to critique the previous session, which will be particularly fun, since I think the speakers from the previous session have fled. But it will give us an opportunity to get some more input, I hope, from the audience and some other views on what we have been talking about.

Our first critiquer is Elizabeth Cook, who is an associate in the Climate, Energy and Pollution Program at the World Resources Institute.

13

CRITIQUE OF SESSION II

Elizabeth Cook

Climate, Energy, and Pollution Program
World Resources Institute
1709 New York Ave. NW, #700
Washington, DC 20006

In hearing the four different perspectives on the Climate Change Action Plan, I think three very important themes emerge. The first is that there is a gap, a gap between the objectives of our efforts and current policies to address climate change. That's true at both the international and the national level.

Second, regarding the U.S. Climate Action Plan, there is a clear consensus that everyone wants to make it work. But the question is: how can we do that?

And lastly, in the long term, it's clear that more fundamental technological changes are needed to adequately address climate change.

This morning we heard about the global picture, the science and the international politics. I think it's worth reviewing a couple points quickly, because it provides an important backdrop for our national effort.

There is a recognition that there's a very large gap, you could say, between the science which drives the policy and what the existing policy now is within the Climate Convention. Firstly, the Convention does not yet address any actions beyond the year 2000. Not all countries have agreed to take actions to reduce greenhouse gas emissions at this point, only industrialized countries. And ongoing negotiations to strengthen the convention, we didn't hear much about this this morning, are very contentious.

The science process which drives these negotiations and guides them is continuing to firm up. The point was clearly made that we do need to stabilize concentrations. It is our objective. What level we need to achieve still needs to be determined, but we know that it is well below a doubling of carbon dioxide.

Watson thought that perhaps it was too soon, or we didn't have enough information to make a judgment, but I think that statement is based on the work of a number of scientists that are looking at the ecological impacts of climate change. And there's no question that under what we have agreed to internationally in the Climate Convention, emissions will continue to rise. There is a very large gap between where we are headed and getting to stabilization at below a doubling.

Well, during this afternoon's session, we heard more specifically about a national gap between what we are trying to achieve in the near term, which is stabilization by 2000, and what the Climate Action Plan is likely to deliver. Mr. Lashof put up some very important

Evaluating Climate Change Action Plans, edited by White
Plenum Press, New York, 1996

information that shows that that gap may be as large as 70 million metric tons of carbon equivalent. That could be due in large part to oil prices, a growing economy, congressional funding which might be delaying implementation of the plans, and accounting schemes, such as using trees to offset carbon dioxide.

Mr. Beck from Edison Electric also mentioned that with utility restructuring, there may be some changes that may even make it difficult for some of the voluntary programs to work.

Well, what can we take from this? I think it is clear that our very first step is very much in the right direction, but that we're finding that it's a more challenging task at hand than we had originally anticipated. And I would add that we're not alone in finding that. We heard a little bit about that today, but we'll hear more tomorrow that a number of other countries are finding a similar challenge in meeting the 2000 goal.

But what's clear from the big picture is that we can't do this alone, but the U.S. is clearly a major emitter of greenhouse gas emissions. So what we do here and how well and effective we do it, will set an example. And, in fact, we will hopefully develop the kinds of technologies that will be used globally to tackle the deeper cuts of greenhouse gas emissions that are needed.

So what did people say today about the Climate Change Action Plan? Well, I think the most impressive message that we heard from everyone was enthusiasm. You heard things about saving money, that it's good for business. This is one of the most quickly implemented Federal program we have ever seen, and these programs are good for economic development at the state level. All those things are very positive.

The question is: How do we address the gap? It's very important, as we know, for the voluntary programs to be effective, because otherwise some of the regulatory options may not be as desirable for the industries engaged. But at the same time, it's not entirely clear that the 44 measures in the Climate Change Action Plan will have enough bite to meet the year 2000 goal. Some of the weaknesses in the Climate Change Action Plan are similar weaknesses that some other industrialized countries may have in the plans that they've submitted. For example, that it doesn't tackle the growing emissions in the transportation sector, necessarily; that many of the plans, including the U.S. plan, have steered away from energy pricing. John Hemphill mentioned that that would be a very useful measure; that there is no enforceable cap, something that would ensure that we keep our emissions at a certain level. Whatever kind of scheme used to implement a cap is not in our plan. It's not in many of the other plans.

So given the political climate that we have, a political climate which might be less inclined in a new Congress to address some of these harder political measures, how can we make the plan work?

Well, I thought Lashof had some very specific suggestions. Most of them dealt with improving efficiency, efficiency for autos, appliances, and homes. John Hemphill talked more broadly about the need for U.S. lending and investments in technology abroad to focus on cleaner energy technologies, and those are other options.

But what I found interesting today was that there wasn't a clear consensus about how important congressional funding is, or, as we've heard, Secretary O'Leary was quoted saying that the funding levels could cripple the program. And Lashof showed what he thought might be some of the problems with less funding. In fact, implementation might be delayed. There was some sense that we could do this without funding. However, I think it's important to note that David Rinebolt mentioned that this funding is what's catalyzing these programs at the state level. We need to be somewhat concerned about the funding for these programs in a new Congress, given that all sorts of programs will be on the chopping block.

I would suggest that we continue to make an effort get these programs fully funded. There is a coalition called the Sustainable Energy Budget Coalition, where environmental organizations are working with industry. In fact, they're working with the Business Council for Sustainable Energy Future. And I would encourage other industries that have been involved to get more involved. In particular, both environmental organizations and industry groups need to be heard at the district level. There's going to be all sorts of new members of Congress needing to learn about these programs, needing to know that voluntary programs can work, that they're useful, that they do encourage economic development, and that they are good for business. It's important that that message be delivered. There's no more potent messenger than to hear that from industry.

If the alternative over time, as both Dan and John alluded to, is some sort of mandatory standards or some sort of energy pricing reform, I think that one opportunity for dealing with these policies might be the fact that other countries in the OECD are having difficulty in meeting their near-term goals.

It was mentioned today that New Zealand has talked about a backstop of a carbon tax and that the Climate Treaty itself offers and encourages countries to coordinate measures, in particular on economic instruments. This could be one way in which countries could get together to talk about some of the competitive or distributional concerns, associated with changing energy prices. They could figure out a way in which they could move forward and evaluate some of these politically more difficult measures, measures that might have the bite that is needed to go beyond the 44 actions in the Action Plan.

At the World Resources Institute we foster that kind of dialogue among governments on both sides of the Atlantic, and with not only policy-makers but non-governmental organizations, too.

So these things may not be politically feasible now as standards or taxes, but there is a possibility that they can be evaluated in a coordinated fashion with OECD countries in the context of the international Climate Change Convention framework.

But beyond carbon dioxide and the energy sector — I have to admit I'm not an expert in energy — there are opportunities near term for reductions in other greenhouse gases. No one directly addressed that today, with the exception of Bob Beck who mentioned the comprehensive approach.

One of the very positive things about the U.S. Climate Change Action Plan is that it takes steps to deal with gases other than carbon dioxide. In fact, it gets roughly one-third of its reductions from dealing with methane, fluorocarbons, and hydrofluorocarbons. There are other opportunities within emissions of those gases that can be addressed in the near term to bring about some additional reductions. I'll point out one that Bob Watson talked about this morning.

Sulphur hexafluoride is going to be cited in the Guinness Book of World Records for global warming potential when the Working Group 1 report is published by the IPCC. That's a gas that we don't know very much about. It's not used very widely, but it is used, for example, in the electrical utility industry as a potent electrical insulator. There are ways in which emissions of that gas can be mitigated, whether it's through recycling of that gas or alternative insulating materials. Perhaps the Edison Electric Institute and the elecrical utility sector should look at sulphur hexafluoride.

So beyond these ideas for the near term, there is the longer term. And how can we deal with those fundamental changes, the steeper cuts that are needed after some of the cost-effective short-term steps are already taken? Some of our speakers did talk about that. Dan talked about the trajectory. John talked about the need to make a broader shift towards sustainable energy. And, in fact, David mentioned that some of these state programs are already initiating those changes.

But I think there needs to be a lot more work done, and it's not clear what the answers are, how we get those new vehicles onto the street and how they get developed. We heard of some efforts through the Federal Advisory Committee, the Super Car Initiative and the

Edison Electric Institute, but it's vital that we initiate a technological shift to more efficient and electric vehicles, whether they're driven by battery or eventually by hydrogen fuel.

The last thing that I'll mention, just to emphasize what John said, that there's a real opportunity for a technology initiative, something that would be an international initiative, that would be coordinated among industrialized countries to target money now going towards renewable energy programs and to pool it and to have a very potent commercialization impact through aggregated strategies. The objective would be to reduce the cost of renewable energy technologies, so that we have cost-effective, non-carbon, renewable energy technologies available to make the kinds of deep cuts in greenhouse gases that we need.

We should pool the resources that are allocated in very disparate parts of the world and direct them to have a more potent impact, both a technology push and market pull effect.

To summarize my statements today, the policies that we now have will not do all that we need to reduce the buildup of greenhouse gases. And I think that point was clearly made by a number of speakers.

But even the existing policies are going to be very challenging to implement. We're seeing examples of that both here in the United States and internationally.

It's vital that there's industry cooperation. It's very encouraging. It's important that governments, NGOs and industry continue to cooperate on both voluntary programs and the long-term changes that are needed. One strategy that we could pursue is that actions that have a bigger bite on these greenhouse gas emissions, but may not be as politically easy to implement, could be more easily addressed through coordination and cooperation with other industrialized countries. The Framework Convention on Climate Change offers that opportunity, both for near-term measures and also longer term goals that will bring about the kinds of technological changes that are needed.

14

CRITIQUE OF SESSION II

David Montgomery

Charles River Associates
1001 Pennsylvania Ave. NW, Suite 750N
Washington, DC 20004

I understand my assignment to be one of trying to recap some of the themes in this afternoon's session and perhaps to critique some of the ideas that were discussed.

Let me start out with an oversimplified reaction. There were four speakers this afternoon, and from that I would infer the states are interested in how much money they can extract from the Department of Energy and have provided some examples of what they have done with it. Industries that will benefit from the climate action plan have emphasized how well they do and tend, in this case, as in others, to ignore losses that will occur elsewhere in the economy when resources are spent on energy efficiency and renewable energy rather than goods and services that consumers would have wanted and would have consumed.

Take electric utilities for example. I would point out that they are pursuing, as part of the climate plan, what makes sense for business, and that they probably would have pursued this anyway. Environmentalists have adopted a somewhat broader perspective, pointing out that the U.S. Climate Action Plan does not appear to meet its goals, and arguing that the U.S. should both adopt more ambitious goals for the year 2000 and should embrace a more stringent regulatory regime in order to ensure that the country meets those goals.

I thought that this session might try to evaluate the U.S. Climate Action Plan and where we stand on it. And that, I think, requires looking at things a bit more broadly. It means asking what is worth doing now, considering the costs and benefits of actions that might be taken, and comparing present opportunities with future opportunities for accomplishing the same thing.

In order to come up with that kind of an assessment, it's really necessary, I think, to begin by examining the U.S. role in the world carbon dioxide and greenhouse gas emission picture and consider a time horizon somewhat beyond the year 2000. And I'd like also to put some of the comments from earlier this afternoon in that perspective.

How important is what the U.S. does by the year 2000 in terms of the global picture of climate change? Figure 1 is a projection of global carbon dioxide emissions taken from the book published by Alan Manne and Richard Richels, which shows, first of all, the long time scales that are involved in any process affecting climate change.

From 1990 global emissions of carbon were something like six billion tons of carbon, growing to perhaps close to forty billion tons of carbon by the year 2100. That's a long period

Evaluating Climate Change Action Plans, edited by White
Plenum Press, New York, 1996

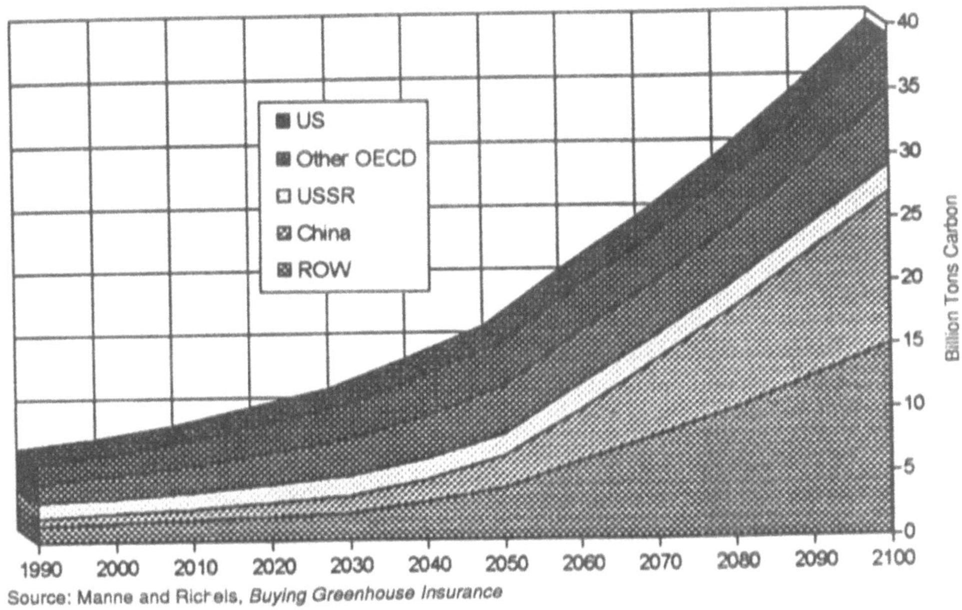

Figure 1. Long-time scales and shared responsibility for emissions – global carbon emissions.

of time. And the growth in emissions between 1990 and 2000 is dwarfed by what is going to be happening in the future.

The U.S. is depicted at the top of figure 1. Below that are emissions from the rest of the OECD. The USSR is the next band, then we come to China and the rest of the world.

Right now the industrial countries are responsible for about half of total carbon dioxide emissions. By the year 2100, the growth in China and the rest of the world will make those countries responsible for perhaps two-thirds of global emissions, i.e., the bulk of the emissions over the next century are going to be coming from these countries, not from the industrial countries, and certainly not from the United States.

The United States and the rest of the industrial world could vanish from the face of the earth, as far as carbon dioxide emissions go, and it would make precious little difference to the total concentrations of carbon dioxide in the atmosphere by the year 2100.

We cannot detect on this chart what difference it makes whether or not the U.S. achieves any specific goal for carbon dioxide emissions over the next five years. This is a problem involving a much longer time horizon; what happens in any given year really is not relevant. It's the whole area on that chart that matters for the accumulation of greenhouse gases in the atmosphere — the perspective that we need to adopt when we discuss any of these action plans and evaluate what makes sense.

Let me make a couple of other points that I would distill from the presentations today.

The first one is that the Climate Change Action Plan does not appear likely to meet its goals, because growth in carbon dioxide emissions from the U.S. is likely to exceed the claimed savings from the plan. That point has been made by several speakers, and I think it's absolutely correct.

The Department of Energy's own projections of carbon emissions to the year 2000 are that they will grow by 145 million tons. That number comes from the Energy Information Administration. The proposed plan itself only claims a reduction of 110 million tons. So there's a gap built in when we start.

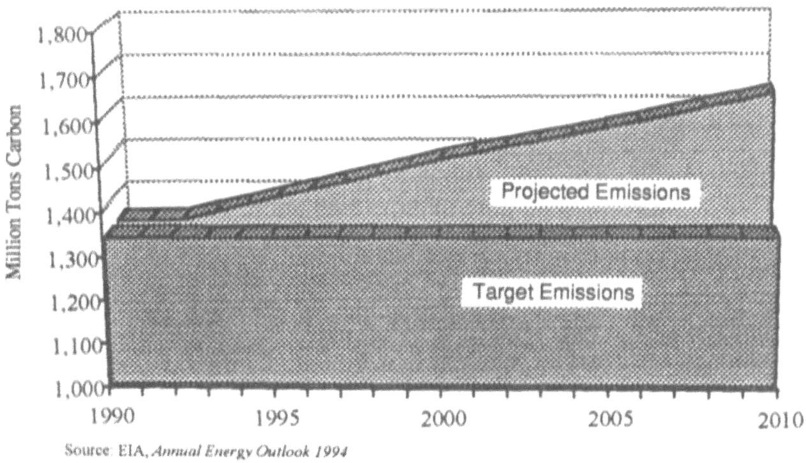

Figure 2. CCAP can't offset 145 MMT growth in emissions by 2000.

In addition, we really don't know what savings will result from the Climate Action Plan. And I think what the speakers have said today actually underscores that fact. We also don't know how much of the savings projected in the plan are actually already in our baseline forecast of energy emissions.

We have a target that is to restrict emissions at today's levels through the year 2000. And just to make it interesting, I've extended figure 2 out to the year 2020. We have a forecast, which I've taken from the Energy Information Administration's Annual Energy Outlook, of emissions growing by about 145 million tons between the year 1990 and the year 2000.

EIA includes in its forecast some very aggressive assumptions about energy conservation and reduction in the amount of electricity that's going to be used per unit of GNP. Thus, this baseline of 145 million tons, which already exceeds what the Climate Action Plan is likely to achieve, already includes a lot of energy conservation measures. If we just count these energy conservation measures when we score the climate action plan, there's going to be a great deal of double counting of what is already included in the baseline forecast.

I think Dan Lashof is absolutely right that the climate action plan is not likely to achieve its goals. It's a plan which is premised on, if you like, "no-regret" strategies, voluntary measures which may make economic sense in their own right, but are not likely to achieve the goals that were stated.

What's wrong with this picture? I submit that there is absolutely nothing wrong with this picture. This is exactly how things ought to be. But it would have helped a great deal if the sensible idea of a voluntary program directed at measures that reduce emissions and are good business had not been mixed up with a set of arbitrary goals that distract attention from the key question: what is it that makes economic sense to do today?

The focus on low-cost or no-cost options also makes perfectly good sense in light of what we actually know about climate change. This go-slow approach is what the National Academy of Sciences recommended several years ago, given the difficulty of identifying or quantifying the economic damage from climate change. It would take convincing estimates of large future damages that could be avoided through climate policy to justify substantial near-term costs. And that's exactly the issue that ought to be debated if we are talking about

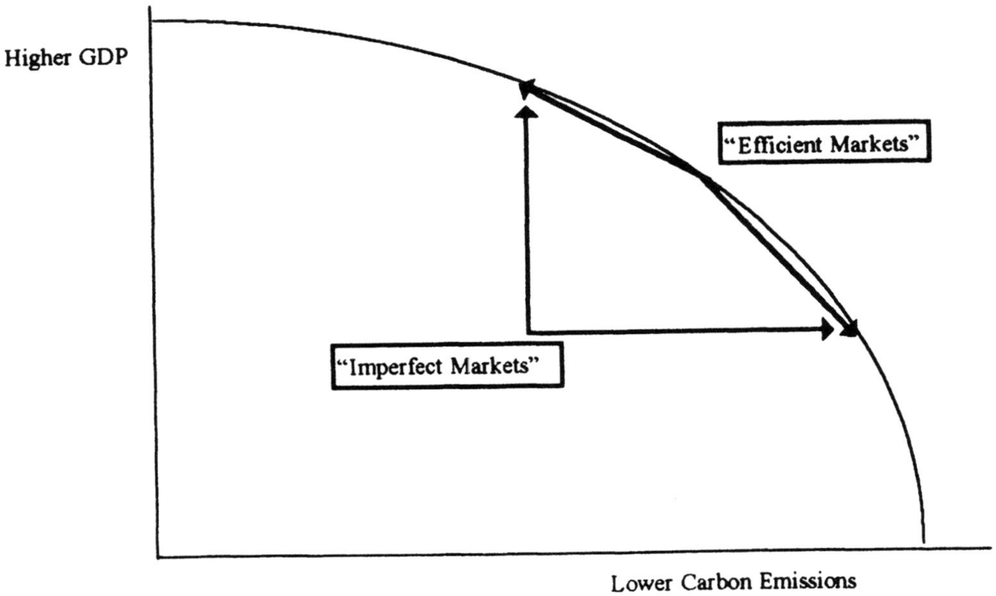

Figure 3. Market imperfections and no cost options.

whether or not to do more. So, what do we know about the damages and what do we have to do today in order to avoid them?

I will stick with the second part. What do we have to do today in order to avoid damages? Let's begin by thinking about this notion of programs with voluntary measures designed to accomplish a "no-regrets" strategy. We shouldn't get hung up on whether or not some arbitrary target for emissions reduction is met if the idea is to address market imperfections in ways that will reduce both costs and emissions.

Now let me pause to repeat a question which I think applies to virtually everything we heard this morning. Why does anyone expect a voluntary program to work and achieve energy conservation that is not already included in the baseline?

I was one of the lead authors for the IPCC second assessment in the volume that is evaluating mitigation policies. We spent a great deal of time considering the issue of "no-regrets" strategy.

What is it that makes it possible to design policies that are good for the economy and reduce emissions at the same time? Basically what it takes is something that economists call market imperfections. I want to try to illustrate what these are first and then talk a bit about why the search for policies that remedy market imperfections does not necessarily provide you with any particular target for emission reductions.

We can think about a tradeoff in the economy, which is expressed by this kind of curve in figure 3 where we have lower carbon emissions on the horizontal axis and a higher GDP on the vertical. If, as economic theory has it, markets will work efficiently, then we're not going to be able to effect any improvement in carbon emissions unless we give something up in terms of economic performance; that is, we have to pay some cost in order to obtain a reduction in carbon emissions.

Likewise, if we tolerated even larger emissions of carbon, some economic gains could be achieved—if, for example, we're already constraining carbon emissions through existing regulatory programs or efficiency standards. So we start at a point on this frontier, and the only way we can get lower emissions is by moving to lower GDP at the same time.

The alternative view is that everything doesn't work perfectly in markets, or that we can find ways in which the market system is not producing an outcome in which we are obtaining the greatest GNP possible or the greatest benefit from our resources. That's a case where we would be inside this frontier, and it would be possible to move in a direction that gives us both an improvement in the economy and a reduction in emissions.

Well, what would that require? It would take some specific things that are not working right in the economy, things that somehow prevent the right price signal from being communicated and acted on when people make decisions about energy production and energy consumption.

What is usually mentioned in this regard are things like lack of information about how infrequently-purchased items will perform, liquidity constraints that make it difficult for some households to finance purchases, or so-called "agency problems" in which decisions about energy use are made by someone other than the person who pays the energy bills. We used to point to master -metered apartments as an example of this, but we found through our energy consumption surveys in the Department of Energy that master-metering pretty much disappeared when the price of electricity started to go up rapidly. So we have lack of information, liquidity constraints, agency problems, and other things that don't work right in the market economy.

It's not clear that voluntary programs, as we normally think of them, address market imperfections, since these programs are ones which work only if it's in the private economic interest of someone to do something. So it shouldn't be surprising if we find that the voluntary programs fail to give us a great deal.

What seems to have been missing in the policy area is a willingness to concentrate on the hard task of studying what these market imperfections actually are and designing policies that could reduce them. An example could be something like the proposal for energy-efficient mortgages. When people finance their houses, or qualify for mortgages, we don't take into account energy costs as part of the overall financing, and this gives the wrong incentives, both for buyers and for builders. Well, perhaps the proposal to back energy-efficient mortgages might alleviate this problem in some ways.

That's a very good idea, and it's a very good way to design policy. But it won't necessarily result in any specific amount of reduction in emissions, because the emission reduction is determined by how many opportunities there are to fix things in the economy, and, in particular, how close the economy is to already being on this frontier.

Any attempt to design policies that deal with market imperfections is not necessarily going to bring about any specific accomplishment in terms of energy savings, because doing all of this is hard. It's hard to identify where the market is not working well, and it's generally impossible to predict what the energy savings are going to be from attempting to ameliorate market imperfections. We really don't know enough, and we have not paid enough attention to identifying what these market imperfections are. Their extent is not known, even if their existence is surmised. And the problem is that policies have to be designed quite carefully to deal with the specific market imperfections.

Efficiency standards and demand-side management are not the answers to every market imperfection. In some cases, they can make the economic problem worse, even if they seem to be moving emission levels in a direction that might be desirable.

It does not appear that the Climate Action Plan is likely to hold emissions to 1990 levels. We shouldn't expect a great deal from a plan which is focusing on voluntary measures and trying to identify market imperfections that could be remedied in ways that would produce both economic benefits and reductions in emissions. But those are all quite legitimate things to do and they make economic sense, but they don't lead to any particular

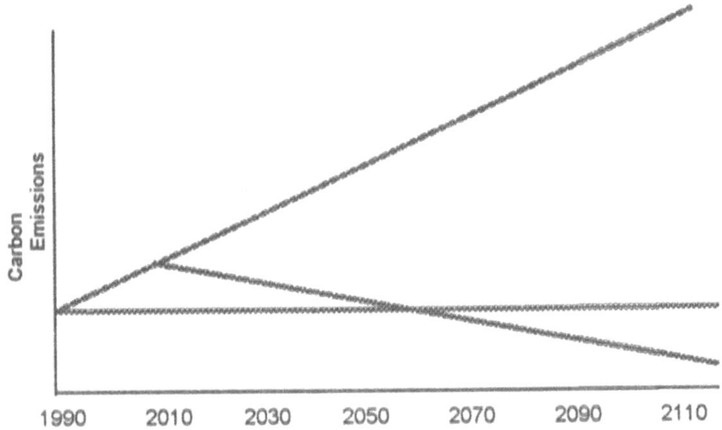

Figure 4. Delay can reduce cost and achieve the same goals.

level of reduction in emissions. They simply lead to the level of reduction which makes economic sense right now.

Now that's just fine when we look at the time scales that are involved in climate change and look at the options that we have for deciding when reductions in emissions should be taken.

Figure 4 is really a stylized chart, where the higher line is baseline growth in carbon emissions. Now, as I mentioned, I think virtually everything that's alleged about the science suggests that it's total concentrations of greenhouse gases in the atmosphere that matters for climate change, and this quantity depends on cumulative emissions over periods of time that are much longer than the remainder of this decade.

I think Mr. Hemphill mentioned the importance of long-term and strategic thinking about climate change issues and policy responses, and that's exactly what we need. I compare here two policies, one of which would be a strict adherence to a limit on carbon emissions at 1990 levels, which is the horizontal line. The other policy allows some increase in emissions for a decade or two, and then reduces emissions down to or below the level of 1990, sometime during the course of the next century.

The lines are carefully drawn so that the total emissions we would get over the next century if we hold emissions at 1990 levels is equal to the total amount of emissions we would get if we let them go up for a few years and then brought them down below 1990 levels later in the century.

There are three reasons why the strategy of letting emissions increase for a while and then bringing them down in the future is much less costly than a policy of adhering to a strict limit on emissions today. Basically, what we would be doing is allowing more time to manage the transition to lower carbon intensity, which would, in turn, lower the costs of emission reduction.

The three reasons for this are fairly simple. The first is that many of the efficiency measures we think of, and certainly most of the measures we think of for substituting non-fossil fuels for fossil fuels, become much cheaper when the capital stock turns over, compared to cases where we try either to retrofit our existing capital stock or to replace the capital stock prematurely. And turnover of that capital stock, especially if it includes structures, can take 30 or 40 years.

Second, there are opportunities for conducting R&D that will make the cost of future non-fossil technologies much lower than the cost of anything we can think of deploying today. In this regard, I think I would take issue with what the first speaker this afternoon said in complimenting the Department of Energy for taking its dollars and shifting them out of R&D and into the deployment of technologies. That policy could be counterproductive and very wasteful in the long run, if the same dollar resources could be leveraged into developing much superior technologies for future application. Because, with regard to climate, it doesn't matter when we remove the emissions, as long as they don't get into the atmosphere over this long period of time.

And finally, there's simply the time value of money, which tells us that we ought to think about whether we want to act today or act tomorrow, especially if it's cheaper to do it tomorrow because of the turnover of capital stock and opportunities for R&D.

Think about the rate of return on investment, and suppose that $100 spent on emission reductions in the year 2000 could reduce emissions by one ton. Suppose instead that we take that $100 and devote it to capital investment and reinvest those returns for 20 years; in that case, we simply leave these resources available for investment in the economy, and the output of the economy will grow at a rate equal to the rate of return on investment for the economy as a whole. Conservatively, that's about six percent real return before taxes, because we're considering what comes back to the economy as a whole - not just from private investors but the share that the government takes as taxes.

If the rate of return on capital is as little as six percent, this $100 investment will grow to $320 after 20 years. That means that we could get 3.2 times as much emission reduction 20 years from now as we could today, simply by investing the resources elsewhere in the economy. We have to think about discounting future costs and comparing the costs today to the discounted present value of costs that we would incur tomorrow, all of which costs are likely to be lower than those involved with trying to take strong actions today, because those future actions will occur naturally as the capital stock turns over. And we will be able to take advantage of new technologies for fuel substitution that are not available today.

Thus, a comprehensive R&D program needs to be a very important component of what we do today and over the long time scales of climate change, it will make a great deal more difference than hitting an arbitrary target for the year 2000.

Outside the R&D area, voluntary programs and a policy development process that identifies and proposes remedies for specific market imperfections both offer a very attractive prospect. When market imperfections are ameliorated through appropriate structural remedies, e.g., things like changing the way banks give out mortgages, as opposed to more command-and-control regulation, there are no losers — at least there are no losers as long as we have a realistic appraisal of the administrative costs of bringing about these changes.

Emission reductions that are achieved in this way can bring about improved economic performance. We just have to abandon the illusion that somehow enough market imperfections can be found and fixed to get the country to some arbitrary goal. R&D can make the transition to lower carbon intensity much less costly, and investment in science can clarify what the benefits of these actions might be. The time scales for the problems to develop and for the solutions to be found are both quite long. These, I think, are the important issues — not whether or not the U.S. will make or miss an arbitrary goal in the next five years.

And I think there are also policy decisions to be made in the sense that Dr. Watson mentioned this morning. Policy decisions are not just how much to spend today or how much to do today to reduce emissions. Policy decisions must be made to assemble a plan that includes R&D and long-term changes in the way we do business.

DISCUSSION

Mr. Chase: This business of a window of opportunity, I would maintain that window is open now and was opened by the Clean Air Act amendments of 1990. Right now utilities are faced with meeting requirements for a reduction in sulphur dioxide. Many of the plants with commitments to meet are very old plants that have been through a period of life extension processes.

So the window of opportunity is not out at 2010 or 2020. The window of opportunity is open now. The real opportunity exists because people are making decisions about repowering, building new plants — we haven't built a new plants in a long time, but push is coming to shove right now.

I would also like to say that I think that cost benefit analysis is inappropriate to this problem. If you are a careful driver, you still buy car insurance. If you are a careful driver, you never have any claims. And the cost benefit analysis would say that you are fool because you've been wasting your money.

The Climate Change problem is much more like driving an automobile. We don't know what's down the road. We don't know how much time we have, because the atmospheric processes are not linear systems. There may be thresholds out there that once we turn the bend, we may be in a real jam. The prudent policy is to take as much action as is politically and economically feasible, as fast as we can. I would agree with you, I don't think we should look at arbitrary goals, but rather the ultimate goal that was mentioned of getting concentrations - not emissions, but concentrations — back to some stabilized level.

Dr. Montgomery: Let me just try to throw out three or four ideas on each of your points.

Certainly, repowering coal-fired power plants that are ending their nominal lives with another fuel is a way of reducing emissions, but it's an expensive one, because what you have to do is compare the cost of continuing to operate those coal-fired power plants or repowering them with coal with the cost of building a new power plant. And when I sit down and do that calculation, it just doesn't come out as being a no-cost measure. It is a feasible decision if you were willing to commit to carbon taxes that were, say, between $100 and $200 a ton. So it's an expensive measure. It's something that could be done, but the cost of what is in essence shutting down an existing plant and replacing it with a brand new one is very expensive.

On the benefit-cost analysis, I think that, as far as the insurance analogy goes, the best insurance policy we could ever buy is investing in R&D that will result in technologies that make fuel substitution possible on a scale that's large enough to actually make a difference to the climate problem.

At some point, we can no longer have any growth in fossil energy use. If we're going to have growth in energy service demand, which we all think we're going to have and which we all want to have, even natural gas use can't grow anymore if we're going to limit carbon emissions. The best insurance policy I can think of is not doing small things today, but is rather investing in the R&D necessary to deploy technologies in the future. And that's something that would recommend a cost-benefit analysis.

On the final point, I think that's something that really needs to be taken up in the context of an integrated assessment activity, similar to what Dr. Granger is going to talk about tomorrow. But when we're talking about marginal policy changes, when we're talking about actual policy measures that produce a specific reduction in emissions, none of those are going to avoid disaster if we think disaster is around the corner, because we have no idea what it takes to avoid disaster.

So, saying that there is a possible disaster lurking doesn't help discriminate among which policies we ought to adopt, because we really can't tell whether any of them are going to avert the disaster or not, by definition. The disaster is something that you can produce under almost any set of assumptions about where emissions go unless we totally stop them. That's really kind of an analysis stopper, rather than an analysis starting point, I think.

Mr. Hay: What kind of commitment should developing countries make and what sort of time line should they follow in reducing greenhouse gases?

Ms. Cook: Well, the thing that has brought developing countries into the climate convention is the prospect of developing more sustainably than the way that they may be headed by adopting older technologies . They want to address climate change, but they have their own national development priorities that are a first order.

I think that there doesn't necessarily have to be a target that they should meet in terms of reductions adopted near term, but there needs to be assurances that they will receive the kind of assistance that has been promised to them to get on a more sustainable path and to have available technologies to meet their energy needs. And of course as we've heard from many people today, the needs are going to grow much greater in the developing countries, in particular the rapidly industrializing countries. And it's that increasing growth and that trajectory that needs to be changed more importantly than setting a specific target. There are some measures in the Convention which promise to provide that assistance, but it's unclear whether or not that assistance will be sufficient solely through the Convention's financial mechanism. That's one of the issues in discussion of joint implementation, that could offer some opportunities in the future.

Mr. Womeldorff: Suppose that all of the developed countries cease emitting greenhouse gases in 2010 — impossible, but just suppose that — we will never stabilize concentrations. The contrary issue, though, is: how is it paid for?

Ms. Cook: Clearly, the near-term actions that need to happen are that we need to show that the Convention works, that the industrialized countries that are responsible for the greenhouse gas emissions, in the atmosphere today, are willing to make fundamental changes to reduce those emissions, that the commitments that have been made to assist these countries with far less resources and far more pressing environmental problems will be forthcoming, and that those kinds of assistance will help them to get on a different emissions path.

I think clearly after those two things are proven, then you can talk about the kinds of targets or firm commitments that will be needed in order to solve the problem that you're talking about, which is that there are going to be growing emissions and that we need to tackle all sources of those emissions.

Mr. Means: I'm in agreement with much of what David said, but I think at some point possibly in disagreement. One is a delay in the date for meeting any particular emissions goal. This probably produces activities that are dysfunctional in terms of long-terms goals. The alternative for that is investment in R&D.

And the case, therefore, if one leaves aside the problem of international treaties, is one of not only delay but putting substantial funds into R&D in large part precisely because of the graph that David showed for the international development of the carbon dioxide emissions.

Conservation conceivably is a role for the advanced countries but conservation is not the solution for China and India. It has to be non-fossil fuels. R&D in alternative energy sources has tremendous international importance because it is the only practical solution for stabilizing carbon dioxide emissions in the developing countries.

There is a danger that the points that you were making can be read as a call for inaction rather than as a call for a different form of action than setting rigid emissions levels for the year 2000.

Dr. Montgomery: I think that is absolutely right. I agree with everything you said, and I think it's quite well put. But there still are problems with figuring how to do this. With regard to R&D, I think that all of our experience suggests that it cannot be entirely left to government investment; we probably are seeing some of it take place in the private sector already.

For example, I would interpret a great deal of what electric utilities are doing today as being a response to an anticipated, though low-probability, carbon tax in the future. The utilities are doing more now in terms of reducing carbon emissions than you could justify on the basis of today's or tomorrow's prices of natural gas and other fuels because they are anticipating that the government will do something.

That's the crux of the matter. Will anyone in the private sector believe, and therefore work effectively toward new technologies, that there is a genuine commitment to policies 30 or 40 years in the future, when we haven't yet seen any energy policies in the United States implemented for more than 4 years? The government can spend money and public-private consortia can spend money on R&D. We've not seen much result from such spending in the past in the energy sector, so there must be a major impetus to stimulate private investment in order to try to get technology to really work.

Figuring out how there can be a commitment to long-term policies is a real problem, because they are what matter. Nobody's going to invest in R&D because of the thought that they can make money in the year 2000 because we had a goal set for that one year alone.

Mr. Hay: There clearly is quite a bit of remaining confusion about how delay can reduce the cost and achieve the same goals concept. I've got a question, for example, that says, "How far do you ideologically take that? Do you wait forever? How do you determine when your enhanced R&D isn't doing the trick and you have to move to something else?"

Dr. Montgomery: Actually, I think the real point is that you need to tie what you're doing today to what you expect to see coming along tomorrow. The economic principle is one we teach people for thinking about non-renewable resources, like the extraction of oil and gas, which says that if we have a fixed amount of a natural resource the price of that resource has to grow over time at a rate that looks something like the rate of interest.

Exactly the same thing here. We have a fixed resource, an atmosphere that can absorb only so much carbon, if we think of our goals in terms of either concentrations or cumulative emissions. We should have incentives of some kind which start at a low level and rise to a level equal to the cost of the new technologies that we want to deploy.

So if we are optimistic, and we think that we can introduce an ethanol or hydrogen economy or solar and wind power for a cost of, say, $200 a metric ton or an equivalent in, say, the year 2050, discount it back at the rate of interest, and that's how much it's worth today to reduce emissions by a ton, maybe $5 a ton.

I'm not saying that we should do nothing nor that delaying forever makes sense. I say anticipate an introduction date for technologies, look at what it would take to get them introduced into the market at that point, convert that amount to a carbon tax, discount it back to today, and you've got a rationale for a program which makes sense over time and that balances what it's worth spending today with what it would be necessary to spend at some point in the future.

Mr. Hay: How should we be motivating the developing countries, China, to take the right steps at this time?

Dr. Montgomery: I think that's an extremely difficult question. We have not had a great deal of success with foreign aid programs that were designed to bring about changes in the countries we support. So something very different needs to be done.

I heard a very interesting comment on joint implementation from someone who had worked for many years in building oil refineries in different parts of the world. He noted, "We do technology transfer all the time." That's what happens when we build an energy facility in another country. We train people to build it, then turn it over, and then they can do the same kind of thing.

It's very hard to get beyond the traditional policy notion that money has to come from the industrialized countries to actually accomplish something.

Mr. Hay: Advocates of voluntary programs commonly imply that such programs require smaller government budgets than mandatory measures, such as efficiency standards. Is there strong empirical evidence to support this general claim? Is it possible that voluntary programs generally demand larger government budgets than mandatory measures?

Ms. Cook: I don't think I have enough experience to specifically answer that question. I know that today David had said that from a state perspective, to do those kinds of programs initially cost more than a regulatory program that required drafting of regulations and implementation.

Mr. Hay: I would offer an opinion that the programs that have come out of the Climate Change Division at EPA are managed in a very lean and mean way, and I think it probably has been demonstrated that they are more cost effective than the traditional government command and control program.

Ms. Cook: Well, I would add one thing. Having been very actively involved in the CFC phaseout where I participated in a number of voluntary programs with industry, it's interesting how the voluntary programs clearly accelerated action;for example, the introduction of recycling equipment for CFCs from car air conditioners into the marketplace, because there were a whole host of things that needed to be worked out technically. The fact that there was voluntary cooperation really did deliver that product more quickly.

But in the end, it's interesting that the industry itself wanted to have regulation requiring the use of that equipment so that there would be a level playing field from the perspective of the guy down the road who wouldn't go out and buy the equipment that a particular service station would have to invest in themselves.

So it can help both. In that case, there was a role both for the voluntary, which I think was very useful, but there was a request by industry to have a regulatory framework to make everyone participate in the process.

Mr. Chase: I'd like to follow that up. Regulation at its best creates an even playing field for market forces to work.

Mr. Hay: I think we talked earlier about market imperfections and whether more effort wasn't needed to identify them. I would certainly agree with that. I would cite from my own point of view the study that AGA did jointly with the renewables and efficiency industries a couple of years ago, the Alternative Energy Future Study. This looked at some of those market imperfections, regulatory barriers that still commonly exist, particularly at the state level, to a free market fuel choice. We still run into this with Clean Air Act implementation, where a given state will simply tell us, "You can't go to the gas plant. You're going to use our local coal."

I think we have identified a lot of areas where we convinced at least ourselves and some other people that elimination of regulatory barriers could make an enormous difference in terms of greenhouse gas emissions and overall economic efficiency.

Ms. Cook: I actually just have a question for David, because I am somewhat still confused and troubled by the statement that it's okay to allow emissions to go up in the near term, as long as you bring them down more steeply in the end, and that it doesn't make any difference.

I was listening to Dr. Watson today and paying close attention to the issue of the long lifetime of greenhouse gases, the inertia of the climate system and the lesson that we learned by allowing concentrations of chlorine and bromine to get to a certain level that triggered an event in the atmosphere that was nonlinear. Also that by putting more emissions into the atmosphere in the near term, you do actually increase your risk in that time period, even though at some point in time, you may reach an equal level of concentrations.

And I'm just not sure how you can suggest that we ought to make a commitment to that risk in the near term irrespective of what costs in the end might be.

Dr. Montgomery: You seem to be suggesting that this is the one risk that we should eliminate despite all the other risks that we are going to leave in place that affect our planet and our lives. I mean, there are plenty of risks to go around. It's necessary, I think, in thinking about any of these problems to balance aversion with the limited resources at our collective disposal.

Once that balancing is done, we should be talking about level of concentrations. I agree with that. But I do think that it is cumulative emissions that matter, not the emissions on any particular date.

Suppose we take one of the normal upper bounds, which is to keep concentrations from going beyond a level which would lead to temperatures beyond what's in the historical record, [since a model can produce instability at any point], but we scour the historical record without observing such a level.

There's plenty of room, given that level of concentrations, to allow emissions to grow for several decades and then bring them down. We're far from being committed to that.

Mr. Hay: Does anyone favor a carbon tax as a way of doing all these good things?

Ms. Cook: Well, clearly, WRI has advanced the idea of getting the price right as key to having successful efforts to reduce greenhouse gas emissions. Whether or not that revenue should be specifically dedicated to certain activities, I think is always going to be a politically difficult choice in the end, because the CFC revenue, for example, never did go to anything dedicated. There was a lot of talk about how it should go to a fund, and you can't be sure that that's going to be happen.

But clearly, there needs to be changes in the investments from the government and private sector to make some of those changes happen. Whether or not the one fund goes to the other is not something that I'm aware we've put on paper as a policy proposal.

Dr. Montgomery: I think a carbon tax is what has been studied in most of the economic assessments of climate change. And at least one of the reasons for that is a shared perception that, of all the ways of going about controlling carbon dioxide emissions, a carbon tax would be the least costly.

But whether or not a carbon tax should be imposed at any particular level depends on your assessment of how much is worth doing about the problem. It is very much worth debating a carbon tax, because it moves us from talking about how much we could accomplish if we weren't willing to spend anything to what really are the relevant questions: How much is worth doing and at what point in time do we have to do it? I think those are the right questions to ask, and it is a very profitable focus in one's mind.

Mr. Rust: Instead of starting your downsloping at 2010, we could start it at 2030 – and of course none of us will be here to worry about it, if it doesn't work.

Dr. Montgomery: I am really not at all convinced that the world will end if we don't do something about this problem. Again, what we really need to consider are the economic and technological tradeoffs inherent to taking action now as opposed to taking action later.

I suggest that you think about these questions in the context of what I presume Dr. Morgan will be presenting tomorrow, because, really, that's what integrated assessment is, an attempt to pull all of this together in a relatively simple quantitative way, which provides the right balance on the options: you can act too soon, and you can act too late, and here are the factors that you have to consider in order to decide when to act.

Mr. Rust: Can you give us some good, hard facts about these new technologies that are going to be developed? Are they be something nebulous, or are they something that you really know something about, that you can attach some numbers to?

Dr. Montgomery: Well, I think that the main thrust behind real technology development is, in fact, that you develop things that you don't yet know, and you do it step by step, exploiting the best opportunities that you can see. The last thing you want to do is focus on the thing that you know exactly how to do, because the only way you ever win in this game is figuring out how to do the things that you don't know how to do.

Given that, I think you're right. A range of opinion on what these new technologies are going to cost and when they're going to be available needs to be made an integral part of the integrated assessment.

Mr. Rust: Why do you think technologies will kick in 2010 to produce the downsloping line?

Dr. Montgomery: I picked 2010 pretty arbitrarily. Studies that were done in the Energy Modeling Forum, work that Richard Richels did with Alan Manne, work that's going on now in the EMF integrated assessment, point to 2030 as a turning point for achieving the same total cumulative emissions at lower cost. The year picked could be 2010. It could be 2030.

One of the obvious things to consider is nuclear power; it does not involve carbon and is deployable on an unlimited scale for generating electricity. And the 2010 to 2030 time range provides for the conversion of electric power plants and for development of a different generation of nuclear technology options that might obviate the acceptability problems nuclear power has today.

Likewise, you could think of development of a biomass-based technology over the same time scale.

Mr. White: We talk about allowing carbon dioxide in the atmosphere to increase until 2030 and then cutting it down rapidly to maintain a finite quantity but at a lower cost. The climate changes will have happened so gradually that we will be accustomed to it.

If we do not take drastic action now, I would expect that there will be very little incentive in 2030 for us to restructure ourselves to go into a very strict procedure for eliminating carbon dioxide. I suggest we would be pretty complacent in 2030. I'll be dead, but I think the world will just get used to it. If we put this off, then we'll never do it.

Mr. Morrissey: There was one statement I was confused about. You mentioned about allowing emissions to be produced, additional emissions to be produced through 2010, but then you said something to the effect that these emissions would not be allowed to affect concentrations. I was unclear about that.

Dr. Montgomery: I didn't say that. The emissions shown on that graph were annual emission rates. This really is a thought experiment. Let's assume that annual emission rates will continue to grow for some period of time, and then those rates will be brought down by changing carbon emissions per unit of energy consumed.

When those rates come down, there is less carbon being released into the atmosphere in each year, but it's the area underneath those lines which tell us the total amount that's in the atmosphere. I may have misspoken in the process of saying that, but that's what I meant.

The good thing is that it's also where the flat line is – the flat line is interesting only because people keep talking about goals of holding emissions at 1990 levels. An integrated assessment might suggest that there's a trajectory of emissions that would grow over time. And even there, it's worth asking: Do we want a straight line that's growing, or do we want to have the same tilt or more – reducing more in the future and less today?

Mr. Hay: Taking the moderator's prerogative on a final comment, I can't let go of Dr. Montgomery's earlier comments on repowering coal-fired power plants. In this country today, with gas prices that you can tie down under long-term contract, it is in many cases, even for coal-fired power plants, the cheapest way to comply with the Clean Air Act is to tear the plant to the ground and build a new gas-fired power plant on the same site. That's not true in every case, but it's true in quite a few cases.

15

DISCUSSION WITH ELIZABETH COOK AND DAVID MONTGOMERY

Mr. Chase: This business of a window of opportunity, I would maintain that window is open now and was opened by the Clean Air Act amendments of 1990. Right now utilities are faced with meeting requirements for a reduction in sulphur dioxide. Many of the plants with commitments to meet are very old plants that have been through a period of life extension processes.

So the window of opportunity is not out at 2010 or 2020. The window of opportunity is open now. The real opportunity exists because people are making decisions about repowering, building new plants – we haven't built a new plants in a long time, but push is coming to shove right now.

I would also like to say that I think that cost benefit analysis is inappropriate to this problem. If you are a careful driver, you still buy car insurance. If you are a careful driver, you never have any claims. And the cost benefit analysis would say that you are fool because you've been wasting your money.

The Climate Change problem is much more like driving an automobile. We don't know what's down the road. We don't know how much time we have, because the atmospheric processes are not linear systems. There may be thresholds out there that once we turn the bend, we may be in a real jam. The prudent policy is to take as much action as is politically and economically feasible, as fast as we can. I would agree with you, I don't think we should look at arbitrary goals, but rather the ultimate goal that was mentioned of getting concentrations - not emissions, but concentrations — back to some stabilized level.

Dr. Montgomery: Let me just try to throw out three or four ideas on each of your points.

Certainly, repowering coal-fired power plants that are ending their nominal lives with another fuel is a way of reducing emissions, but it's an expensive one, because what you have to do is compare the cost of continuing to operate those coal-fired power plants or repowering them with coal with the cost of building a new power plant. And when I sit down and do that calculation, it just doesn't come out as being a no-cost measure. It is a feasible decision if you were willing to commit to carbon taxes that were, say, between \$100 and \$200 a ton. So it's an expensive measure. It's something that could be done, but the cost of what is in essence shutting down an existing plant and replacing it with a brand new one is very expensive.

On the benefit-cost analysis, I think that, as far as the insurance analogy goes, the best insurance policy we could ever buy is investing in R&D that will result in technologies that make fuel substitution possible on a scale that's large enough to actually make a difference to the climate problem.

At some point, we can no longer have any growth in fossil energy use. If we're going to have growth in energy service demand, which we all think we're going to have and which we all want to have, even natural gas use can't grow anymore if we're going to limit carbon emissions. The best insurance policy I can think of is not doing small things today, but is rather investing in the R&D necessary to deploy technologies in the future. And that's something that would recommend a cost-benefit analysis.

On the final point, I think that's something that really needs to be taken up in the context of an integrated assessment activity, similar to what Dr. Granger Morgan is going to talk about tomorrow. But when we're talking about marginal policy changes, when we're talking about actual policy measures that produce a specific reduction in emissions, none of those are going to avoid disaster if we think disaster is around the corner, because we have no idea what it takes to avoid disaster.

So saying that there is a possible disaster lurking doesn't help discriminate among which policies we ought to adopt, because we really can't tell whether any of them are going to avert the disaster or not, by definition. The disaster is something that you can produce under almost any set of assumptions about where emissions go unless we totally stop them. That's really kind of an analysis stopper, rather than an analysis starting point, I think.

Mr. Hay: What kind of commitment should developing countries make and what sort of time line should they follow in reducing greenhouse gases?

Ms. Cook: Well, the thing that has brought developing countries into the climate convention is the prospect of developing more sustainably than the way that they may be headed by adopting older technologies. They want to address climate change, but they have their own national development priorities that are a first order.

I think that there doesn't necessarily have to be a target that they should meet in terms of reductions adopted near term, but there needs to be assurances that they will receive the kind of assistance that has been promised to them to get on a more sustainable path and to have available technologies to meet their energy needs. And of course as we've heard from many people today, the needs are going to grow much greater in the developing countries, in particular the rapidly industrializing countries. And it's that increasing growth and that trajectory that needs to be changed more importantly than setting a specific target. There are some measures in the Convention which promise to provide that assistance, but it's unclear whether or not that assistance will be sufficient solely through the Convention's financial mechanism. That's one of the issues in discussion of joint implementation, that could offer some opportunities in the future.

Mr. Womeldorff: Suppose that all of the developed countries cease emitting greenhouse gases in 2010 — impossible, but just suppose that — we will never stabilize concentrations. The contrary issue, though, is: how is it paid for?

Ms. Cook: Clearly, the near-term actions that need to happen are that we need to show that the Convention works, that the industrialized countries that are responsible for the greenhouse gas emissions,in the atmosphere today, are willing to make fundamental changes to reduce those emissions, that the commitments that have been made to assist these countries with far less resources and far more pressing environmental problems will be forthcoming, and that those kinds of assistance will help them to get on a different emissions path.

I think clearly after those two things are proven, then you can talk about the kinds of targets or firm commitments that will be needed in order to solve the problem that you're talking about, which is that there are going to be growing emissions and that we need to tackle all sources of those emissions.

Mr. Means: I'm in agreement with much of what David said, but I think at some point possibly in disagreement. One is a delay in the date for meeting any particular emissions goal. This probably produces activities that are dysfunctional in terms of long-terms goals. The alternative for that is investment in R&D.

And the case, therefore, if one leaves aside the problem of international treaties, is one of not only delay but putting substantial funds into R&D in large part precisely because of the graph that David showed for the international development of the carbon dioxide emissions.

Conservation conceivably is a role for the advanced countries but conservation is not the solution for China and India. It has to be non-fossil fuels. R&D in alternative energy sources has tremendous international importance because it is the only practical solution for stabilizing carbon dioxide emissions in the developing countries.

There is a danger that the points that you were making can be read as a call for inaction rather than as a call for a different form of action than setting rigid emissions levels for the year 2000.

Dr. Montgomery: I think that is absolutely right. I agree with everything you said, and I think it's quite well put. But there still are problems with figuring how to do this. With regard to R&D, I think that all of our experience suggests that it cannot be entirely left to government investment; we probably are seeing some of it take place in the private sector already.

For example, I would interpret a great deal of what electric utilities are doing today as being a response to an anticipated, though low-probability, carbon tax in the future. The utilities are doing more now in terms of reducing carbon emissions than you could justify on the basis of today's or tomorrow's prices of natural gas and other fuels because they are anticipating that the government will do something.

That's the crux of the matter. Will anyone in the private sector believe, and therefore work effectively toward new technologies, that there is a genuine commitment to policies 30 or 40 years in the future; when we haven't yet seen any energy policies in the United States implemented for more than 4 years? The government can spend money, public-private consortia can spend money on R&D. We've not seen much result from such spending in the past in the energy sector, so there must be a major impetus to stimulate private investment in order to try to get technology to really work.

Figuring out how there can be a commitment to long-term policies is a real problem, because they are what matter. Nobody's going to invest in R&D because of the thought that they can make money in the year 2000 because we had a goal set for that one year alone.

Mr. Hay: There clearly is quite a bit of remaining confusion about how delay can reduce the cost and achieve the same goals concept. I've got a question, for example, that says, "How far do you ideologically take that? Do you wait forever? How do you determine when your enhanced R&D isn't doing the trick and you have to move to something else?"

Dr. Montgomery: Actually, I think the real point is that you need to tie what you're doing today to what you expect to see coming along tomorrow. The economic principle is one we teach people for thinking about non-renewable resources, like the extraction of oil and gas, which says that if we have a fixed amount of a natural resource the price of that resource has to grow over time at a rate that looks something like the rate of interest.

Exactly the same thing here. We have a fixed resource, an atmosphere that can absorb only so much carbon, if we think of our goals in terms of either concentrations or cumulative emissions. We should have incentives of some kind which start at a low level and rise to a level equal to the cost of the new technologies that we want to deploy.

So if we are optimistic, and we think that we can introduce an ethanol or hydrogen economy or solar and wind power for a cost of, say, $200 a metric ton or an equivalent in, say, the year 2050, discount it back at the rate of interest, and that's how much it's worth today to reduce emissions by a ton, maybe $5 a ton.

I'm not saying that we should do nothing nor that delaying forever makes sense. I say anticipate an introduction date for technologies, look at what it would take to get them introduced into the market at that point, convert that amount to a carbon tax, discount it back to today, and you've got a rationale for a program which makes sense over time and that balances what it's worth spending today with what it would be necessary to spend at some point in the future.

Mr. Hay: How should we be motivating the developing countries, China, to take the right steps at this time?

Dr. Montgomery: I think that's an extremely difficult question. We have not had a great deal of success with foreign aid programs that were designed to bring about changes in the countries we support. So something very different needs to be done.

I heard a very interesting comment on joint implementation from someone who had worked for many years in building oil refineries in different parts of the world. He noted, "We do technology transfer all the time." That's what happens when we build an energy facility in another country. We train people to build it, then turn it over, and then they can do the same kind of thing.

It's very hard to get beyond the traditional policy notion that money has to come from the industrialized countries to actually accomplish something.

Mr. Hay: Advocates of voluntary programs commonly imply that such programs require smaller government budgets than mandatory measures, such as efficiency standards. Is there strong empirical evidence to support this general claim? Is it possible that voluntary programs generally demand larger government budgets than mandatory measures?

Ms. Cook: I don't think I have enough experience to specifically answer that question. I know that today David had said that from a state perspective, to do those kinds of programs initially cost more than a regulatory program that required drafting of regulations and implementation.

Mr. Hay: I would offer an opinion that the programs that have come out of the Climate Change Division at EPA are managed in a very lean and mean way, and I think it probably has been demonstrated that they are more cost effective than the traditional government command and control program.

Ms. Cook: Well, I would add one thing. Having been very actively involved in the CFC phaseout where I participated in a number of voluntary programs with industry, it's interesting how the voluntary programs clearly accelerated action; for example, the introduction of recycling equipment for CFCs from car air conditioners into the marketplace, because there were a whole host of things that needed to be worked out technically. The fact that there was voluntary cooperation really did deliver that product more quickly.

But in the end, it's interesting that the industry itself wanted to have regulation requiring the use of that equipment so that there would be a level playing field from the perspective of the guy down the road who wouldn't go out and buy the equipment that a particular service station would have to invest in themselves.

So it can help both. In that case, there was a role both for the voluntary, which I think was very useful, but there was a request by industry to have a regulatory framework to make everyone participate in the process.

Mr. Chase: I'd like to follow that up. Regulation at its best creates an even playing field for market forces to work.

Mr. Hay: I think we talked earlier about market imperfections and whether more effort wasn't needed to identify them. I would certainly agree with that. I would cite from my own point of view the study that AGA did jointly with the renewables and efficiency industries a couple of years ago, the Alternative Energy Future Study. This looked at some of those market imperfections, regulatory barriers that still commonly exist, particularly at the state level, to a free market fuel choice. We still run into this with Clean Air Act implementation, where a given state will simply tell us, "You can't go to the gas plant. You're going to use our local coal."

I think we have identified a lot of areas where we convinced at least ourselves and some other people that elimination of regulatory barriers could make an enormous difference in terms of greenhouse gas emissions and overall economic efficiency.

Ms. Cook: I actually just have a question for David, because I am somewhat still confused and troubled by the statement that it's okay to allow emissions to go up in the near term, as long as you bring them down more steeply in the end, and that it doesn't make any difference.

I was listening to Dr. Watson today and paying close attention to the issue of the long lifetime of greenhouse gases, the inertia of the climate system and the lesson that we learned by allowing concentrations of chlorine and bromine to get to a certain level that triggered an event in the atmosphere that was nonlinear. Also that by putting more emissions into the atmosphere in the near term, you do actually increase your risk in that time period, even though at some point in time, you may reach an equal level of concentrations.

And I'm just not sure how you can suggest that we ought to make a commitment to that risk in the near term irrespective of what costs in the end might be.

Dr. Montgomery: You seem to be suggesting that this is the one risk that we should eliminate despite all the other risks that we are going to leave in place that affect our planet and our lives. I mean, there are plenty of risks to go around. It's necessary, I think, in thinking about any of these problems to balance aversion with the limited resources at our collective disposal.

Once that balancing is done, we should be talking about level of concentrations. I agree with that. But I do think that it is cumulative emissions that matter, not the emissions on any particular date.

Suppose we take one of the normal upper bounds, which is to keep concentrations from going beyond a level which would lead to temperatures beyond what's in the historical record, [since a model can produce instability at any point], but we scour the historical record without observing such a level.

There's plenty of room, given that level of concentrations, to allow emissions to grow for several decades and then bring them down. We're far from being committed to that.

Mr. Hay: Does anyone favor a carbon tax as a way of doing all these good things?

Ms. Cook: Well, clearly, WRI has advanced the idea of getting the price right as key to having successful efforts to reduce greenhouse gas emissions. Whether or not that revenue should be specifically dedicated to certain activities, I think is always going to be a politically difficult choice in the end, because the CFC revenue, for example, never did go to anything dedicated. There was a lot of talk about how it should go to a fund, and you can't be sure that that's going to be happen.

But clearly, there needs to be changes in the investments from the government and private sector to make some of those changes happen. Whether or not the one fund goes to the other is not something that I'm aware we've put on paper as a policy proposal.

Dr. Montgomery: I think a carbon tax is what has been studied in most of the economic assessments of climate change. And at least one of the reasons for that is a shared perception that, of all the ways of going about controlling carbon dioxide emissions,a carbon tax would be the least costly.

But whether or not a carbon tax should be imposed at any particular level depends on your assessment of how much is worth doing about the problem. But it is very much worth debating a carbon tax, because it moves us from talking about how much we could accomplish if we weren't willing to spend anything to what really are the relevant questions: How much is worth doing and at what point in time do we have to do it? I think those are the right questions to ask, and it is a very profitable focus in one's mind.

Mr. Rust: Instead of starting your downsloping at 2010, we could start it at 2030 — and of course none of us will be here to worry about it, if it doesn't work.

Dr. Montgomery: I am really not at all convinced that the world will end if we don't do something about this problem. Again, what we really need to consider are the economic and technological tradeoffs inherent to taking action now as opposed to taking action later.

I suggest that you think about these questions in the context of what I presume Dr. Morgan will be presenting tomorrow, because, really, that's what integrated assessment is, an attempt to pull all of this together in a relatively simple quantitative way, which provides the right balance on the options: you can act too soon, and you can act too late, and here are the factors that you have to consider in order to decide when to act.

Mr. Rust: Can you give us some good, hard facts about these new technologies that are going to be developed? Are they be something nebulous, or are they something that you really know something about, that you can attach some numbers to?

Dr. Montgomery: Well, I think that the main thrust behind real technology development is, in fact, that you develop things that you don't yet know, and you do it step by step, exploiting the best opportunities that you can see. The last thing you want to do is focus on the thing that you know exactly how to do, because the only way you ever win in this game is figuring out how to do the things that you don't know how to do.

Given that, I think you're right. A range of opinion on what these new technologies are going to cost and when they're going to be available needs to be made an integral part of the integrated assessment.

Mr. Rust: Why do you think technologies will kick in by 2010 to produce the downsloping line?

Dr. Montgomery: I picked 2010 pretty arbitrarily. Studies that were done in the Energy Modeling Forum, work that Richard Richels did with Alan Manne, work that's going on now in the EMF integrated assessment, point to 2030 as a turning point for achieving the same total cumulative emissions at lower cost. The year picked could be 2010. It could be 2030.

One of the obvious things to consider is nuclear power, it does not involve carbon and is deployable on an unlimited scale for generating electricity. And the 2010 to 2030 time range provides for the conversion of electric power plants and for development of a different generation of nuclear technology options that might obviate the acceptability problems nuclear power has today.

Likewise, you could think of development of a biomass-based technology over the same time scale.

Mr. White: We talk about allowing carbon dioxide in the atmosphere to increase until 2030 and then cutting it down rapidly to maintain a finite quantity but at a lower cost. The climate changes will have happened so gradually that we will be accustomed to it.

If we do not take drastic action now, I would expect that there will be very little incentive in 2030 for us to restructure ourselves to go into a very strict procedure for eliminating carbon dioxide. I suggest we would be pretty complacent in 2030. I'll be dead, but I think the world will just get used to it. If we put this off, then we'll never do it.

Mr. Morrissey: There was one statement I was confused about. You mentioned about allowing emissions to be produced, additional emissions to be produced through 2010, but then you said something to the effect that these emissions would not be allowed to affect concentrations. I was unclear about that.

Dr. Montgomery: I didn't say that. The emissions shown on that graph were annual emission rates. This really is a thought experiment. Let's assume that annual emission rates will continue to grow for some period of time, and then those rates will be brought down by changing carbon emissions per unit of energy consumed.

When those rates come down, there is less carbon being released into the atmosphere in each year, but it's the area underneath those lines which tell us the total amount that's in the atmosphere. I may have misspoken in the process of saying that, but that's what I meant.

The good thing is that it's also where the flat line is — the flat line is interesting only because people keep talking about goals of holding emissions at 1990 levels. An integrated assessment might suggest that there's a trajectory of emissions that would grow over time. And even there, it's worth asking: Do we want a straight line that's growing, or do we want to have the same tilt or more — reducing more in the future and less today?

Mr. Hay: Taking the moderator's prerogative on a final comment, I can't let go of Dr. Montgomery's earlier comments on repowering coal-fired power plants. In this country today, with gas prices that you can tie down under long-term contract, it is in many cases, even for coal-fired power plants, the cheapest way to comply with the Clean Air Act is to tear the plant to the ground and build a new gas-fired power plant on the same site. That's not true in every case, but it's true in quite a few cases.

16

ELECTRIC POWER SUPPLY ALTERNATIVES

CO_2 Mitigations

Charles Siebenthal

Electric Power Research Institute
3412 Hillview Avenue
P.O. Box 10412
Palo Alto, California 94303

Yesterday someone said—and I am going to paraphrase—that after the policy makers are finished deciding what has to be done, then it is left to another group of people to decide how to do it, and I guess I represent that end of the equation here today.

I am not going to be able to tell you how the individual utilities who have entered into the voluntary reduction program are planning to achieve those reductions. None of those utilities, to my knowledge, have made that public, and I don't have first hand knowledge of any of their plans. So, it would be inappropriate for me to try to tell you what individual utilities will do.

EPRI has, for about four years, been looking at what actions are available to the U.S. utility industry in the event that carbon reduction of some type is required of the industry. Today I am going to try to share with you the results of those studies and let you know what we have told our members as to what strategies might be available to them.

Figure 1 is a graph which shows one simple thing, and that is: during the last 20 years, during the decades of the 1970s and 1980s, the production of carbon dioxide by our industry per kilowatt hours—that number on the left is millions tons of CO_2 per billion kilowatt hours—has gone steadily down. Unfortunately, that is almost entirely due to the large amount of nuclear power plant capacity which the industry built in the early 1970s and 1980s.

Figure 2 illustrates the long-term trend in the thermal efficiency of the average fossil-fired steam plant. In the years after World War II, we achieved a significant amount of improvement, simply because we were allowed to build new power plants following the war, and new technologies were brought in.

But during the last two decades, the thermal efficiency has held fairly constant. That doesn't mean we didn't improve the thermal efficiency, but we lost all those improvements due to the environmental controls for sulfur and for particulates which were added onto those power plants.

Why is that a problem? Because national statistics show that coal, in terms of the total amount of electrical energy produced in the United States, is over half of the equation.

Evaluating Climate Change Action Plans, edited by White
Plenum Press, New York, 1996

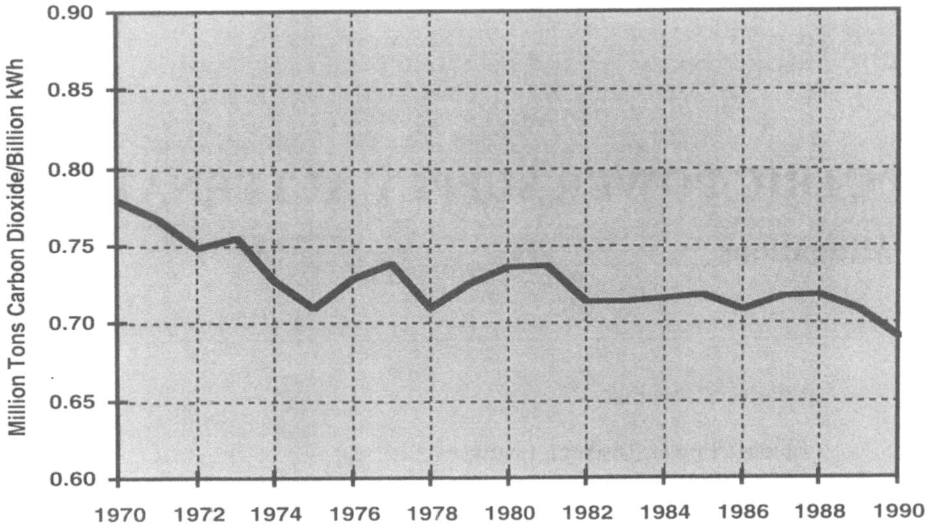

Figure 1. U.S. utility carbon dioxide emissions per total kWh generated.

However, while these national statistics are very useful in letting you know the nature of the problem, they are not representative of any particular utility. For instance, two of the largest utilities in this country, both of which are in California, use virtually no coal at all. So, most of the modeling data that you see, most of the statistics like this for the industry, are very good at defining a national problem, but very poor in terms of trying to figure out what an individual utility might be able to do to respond.

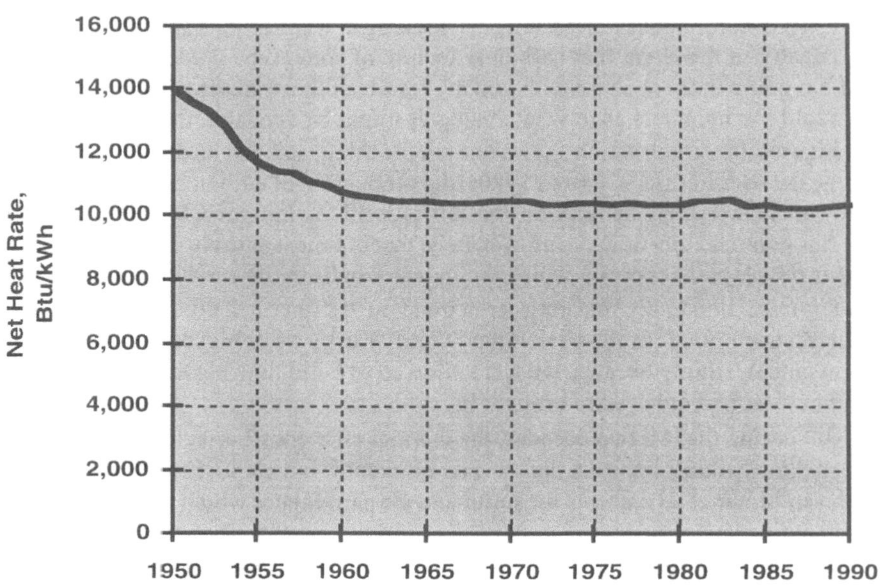

Figure 2. Heat rates for fossil-fueled steam-electric plants.

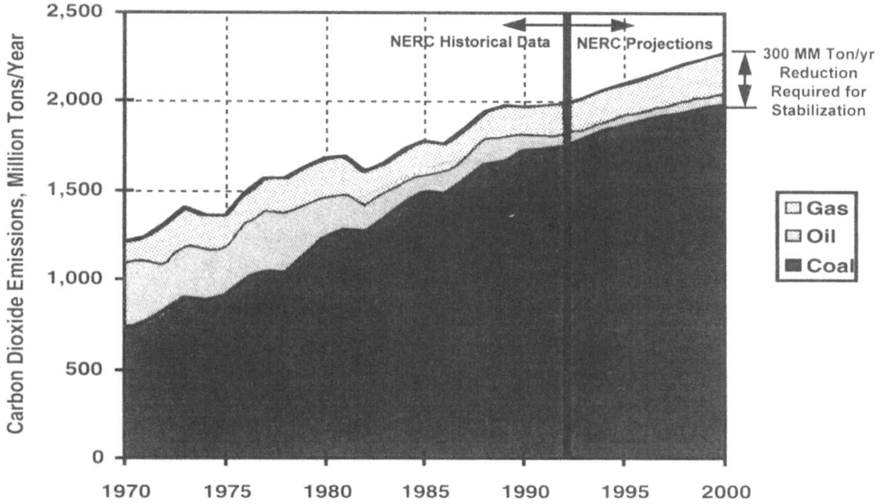

Figure 3. U.S. utility carbon dioxide emissions.

Figure 3 quantifies the problem. It shows utility carbon dioxide emissions for the past 20 years by fuel type, and a projection for the next 10 years, based on the model which NERC uses to predict energy demand and energy responses in this country.

You can see that a business-as-usual policy by the industry with new capacity replacing old capacity using the same fuel mixes, the same technology mixes that we have been using in the past, will lead to a gradual increase in the amount of CO_2 emissions. The question, of course, is what are we going to do about that, both as an industry and as a country, because it really is a national problem. We generate the CO_2 to produce electricity but that electricity is really in response to the national economic demand for energy. And so, it truly is a problem where we must all share a role in the solution.

One of the reasons the electric utility industry has not jumped up and said, "Here is what we are going to do," is the confusion over what the end result has to be. Four potential solutions were mentioned yesterday—controlled growth, stabilization of emissions, stabilization of atmospheric conditions, and then in the extreme, reduction in atmospheric concentrations. Each of those different targets has very different implications over the long-term in terms of the utility industry's strategies.

Turning to figure 4, the top line shows what would happen to CO_2 emissions by one model if we maintained coal in its current proportion and did not introduce anything novel except better coal-fired technologies.

The middle line shows what would happen if, in the year 2000, we stopped building coal plants and just used natural gas to meet capacity addition requirements and to replace retired coal plants.

The bottom line shows what would happen in the future if, in the year 2000, legislation was passed which did not allow us to use coal at all for new plants, and we only used what we call non-emitters, which are nuclear and renewables power plants. The data indicates that the only way the country can achieve a reduction of emissions from electric power plants over the long-term is to stop using fossil fuels, provided that the demand for electricity remains constant or grows in time. While future demand growth is the subject of debate, history shows us that if we are going to have GNP growth in this country, then in all probability we will continue to have electricity demand growth. More importantly, if carbon

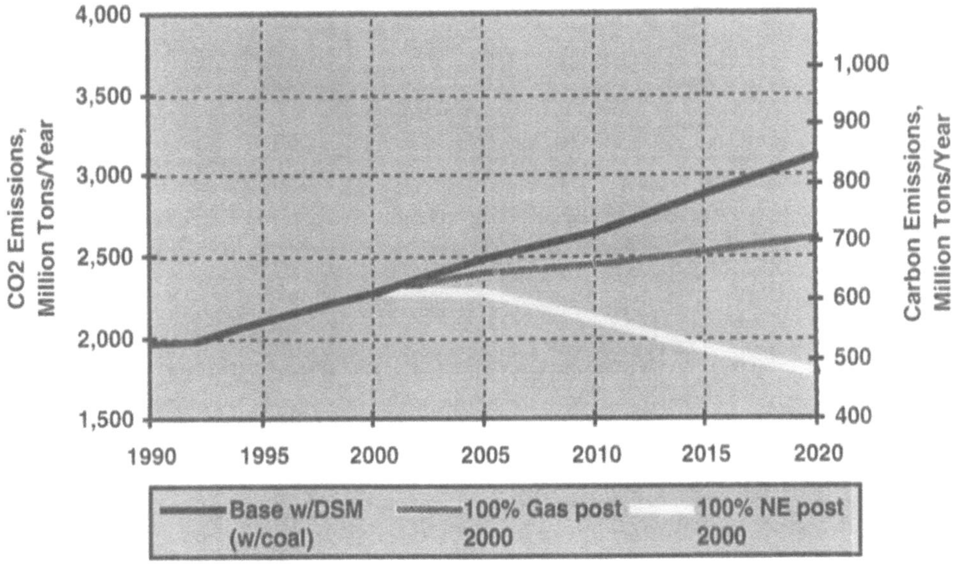

Figure 4. Summary of utility CO_2 emission rates effect of different generation technology additions.

emission restrictions on non-electric uses of fossil fuels such as in transportation, force the substitution of electricity, clearly the electrical component of GNP will become larger.

So, the message of this graph is very simple: if a utility decides to build a fossil-fired power plant today and 10 years from now it is decided that the plant shouldn't operate, all of a sudden it is stuck with a 30-year investment, of which only 10 years have been recovered. Those kinds of concerns are causing a great deal of confusion and indecision within our industry, as to what to do.

I am going to talk first about some of the things that we can do which don't involve capital investment. They are called "environmental dispatch" and "demand-side management." The latter one I am sure you have heard of; the first one you probably don't know a great deal about. Before I can tell you about environmental dispatch, I have to make sure that you understand how utilities operate their generation units.

A typical utility dispatcher sits in his office with a portfolio of power plants to dispatch to meet electrical demands. He does that by a commitment schedule which is pre-planned for the day. It is customer demand driven, and it basically dispatches the units in an order inverse to their cost of electricity production which is controlled by their thermal efficiency, the price of fuel, and the cost of starting up and shutting plants down. The dispatcher's objective is to minimize the cost of generating electricity over the day.

Figure 5 shows a hypothetical economic dispatch schedule. At the bottom of the figure is the hours of the day. It starts at four o'clock in the morning, and this dispatcher's day goes to four o'clock in the morning of the next day.

The shaded area represents the utility customers' demand for electricity. It starts at a very low level at four o'clock in the morning, goes up to a peak sometime around four or five o'clock in the afternoon—I have assumed an air conditioning load here—and then falls off during the remainder of the day.

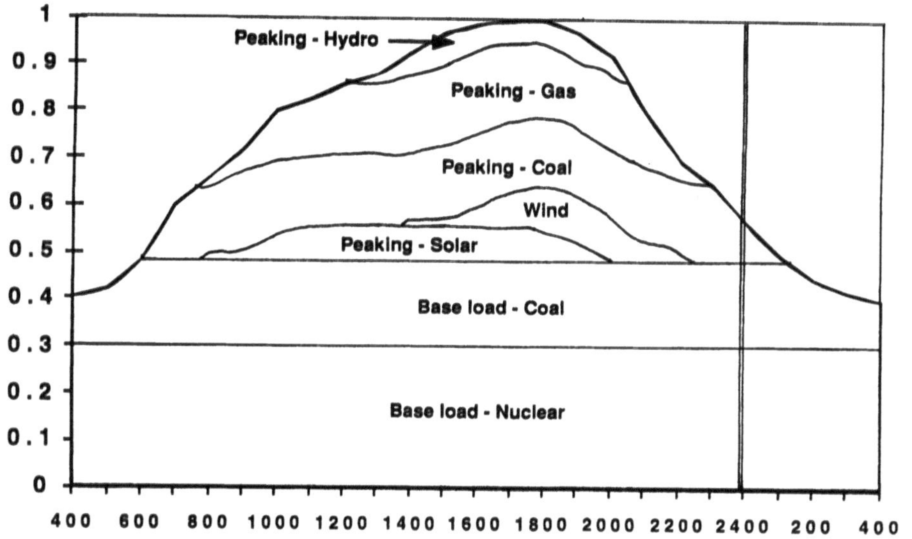

Figure 5. Economic dispatch.

The typical dispatcher has all of his nuclear plants running base load at night, 24 hours a day, 7 days a week, because they have the lowest variable operating cost. About half of the base load coal is in operation at four o'clock in the morning, and as the load grows he increases use of those base load coal plants because they have the second lowest cost.

In this particular utility, I have assumed is located in California, even though we don't have any coal! Sometime around ten o'clock in the morning, some of the solar capacity can be brought on line, because the sun is up and those units are now generating electricity. And again, at least in the San Francisco area, in the early afternoon, the wind comes up and we can generate electricity from wind.

In the intervening period, the peaking coal capacity—that is, that coal capacity which is old, has poor thermal rates, and is too expensive to operate as base load, is used as filler. When we run out of peaking coal, we bring on the gas units—because gas has the highest fuel price. When we run out of gas, we finally use our hydro capacity which in this model, we conserve as a backup to solar wind.

So, what you have here is a very complex way of responding, all based on trying to minimize the cost to the consumer. It is both complex and dynamic. It can vary daily. Some days we miss the guess on how much electricity we need. Some days a plant breaks down and we have to figure out a way to replace it.

Figure 6 shows the basic problem with integrating renewables–that is, non-dispatchable renewables, like solar and wind into the dispatch order. It is that when the sun doesn't shine and the wind doesn't blow, somewhere you have to have a power plant standing by to supply the missing power. That is one of the biggest problems we have in large scale use of non-dispatchable renewables on a utility system. Not only do you have to invest the capital for the renewables unit, but you have to invest additional capital for the back-up power plants to provide electricity when the renewables don't operate. This makes the system more expensive and complex to operate on an annual basis.

These are the conditions which dictate how the supply portfolio is utilized, and the national statistics confirm this. Annual capacity factor data for the nation's generation fleet show that nuclear is utilized almost completely because it has very low operating costs. It is

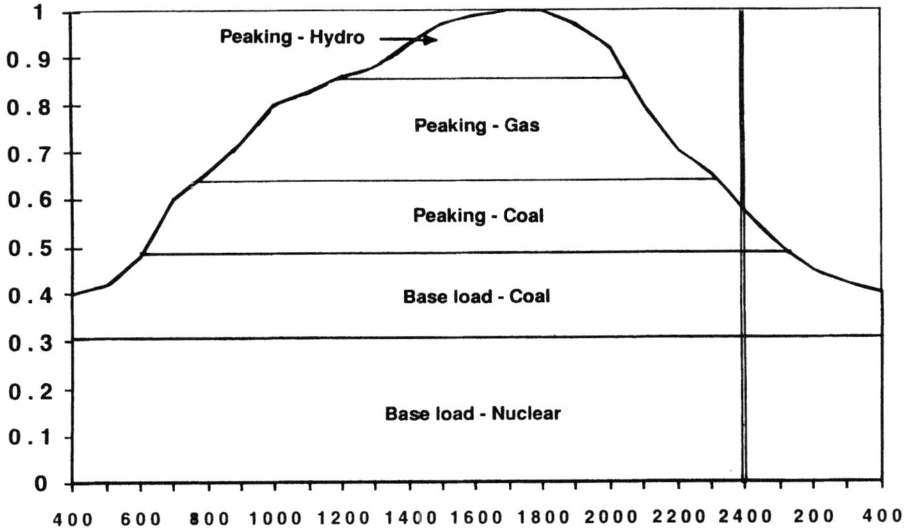

Figure 6. Economic dispatch.

the most economic if you are just dispatching electricity on the basis of variable cost of generation.

Coal plants are used to about three-quarters of their potential, whereas the oil and gas is used to about 30 percent of its potential because they have a much higher fuel price relative to coal and nuclear.

The hydro capacity is only used at 50 percent and that may seem strange to you because hydro has essentially a zero dispatch cost, but hydro is the backup. You need to save

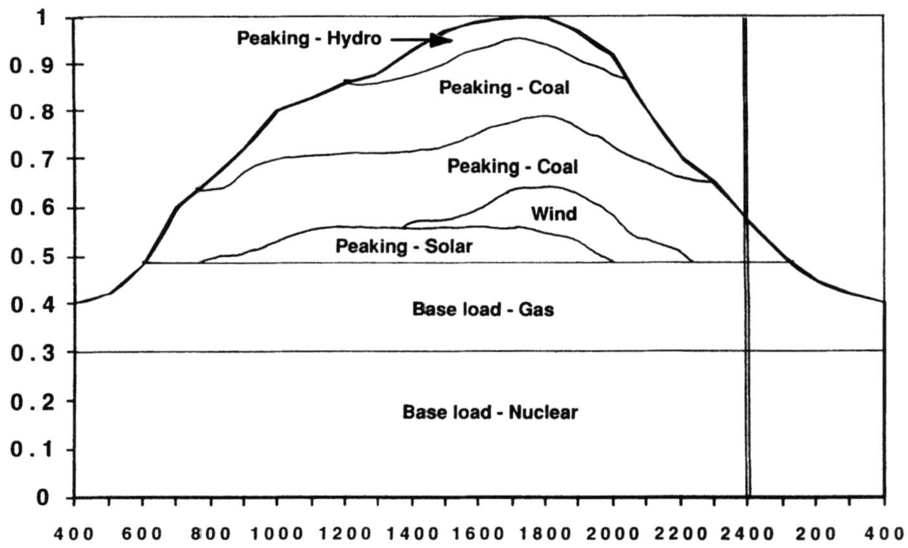

Figure 7. Environmental dispatch.

your hydro resources for when you need them, for the maximum peaking and for when you lose other kinds of power plants. So, it has a high value as stored energy.

Now that you understand the basic concepts of economic dispatch, let's consider environmental dispatch which is simply running the utility system in a slightly different way. In this strategy, we solve the same economic dispatch problem but subject to limits on the total daily emissions of NO_X, SO_X and CO_2.

The results are not surprising. You maximize nuclear use because they are zero emitters. In fact if you can manage to get them to run a larger fraction of the time, you do so, even if you have to invest a bit more money, because each extra kilowatt hour you squeeze out of a nuclear unit produces nothing in the way of air emissions. However, the major effect is to switch the roles of gas and coal. You no longer use gas as a peaking fuel, you uşe it as a base load fuel. Now, that is not free. It costs you money in terms of fuel price.

Clearly, not every utility can do this because not every utility has a lot of gas. In fact, the two California utilities which I talked about earlier cannot utilize environmental dispatch to any large degree because they are as green as they can be. They are already very high on nuclear utilization, they have a lot of renewables, the rest is gas. There is little coal to back out of the mix. So, this kind of thing doesn't work for every utility. But for those utilities that have the right blend of gas and coal-fired capacity, and who can negotiate the gas supply contracts to invert their fuel balance, can use something like this to make marginal emissions improvements without mourning a significant capital cost.

Demand-side management is something that a number of speakers have mentioned. DSM does several advantageous things. I want to show you, although this graph is screwed up, what it really does for the utility system-wide. DSM reduces total electricity use so the utility does not need as much capacity. But just importantly, many DSM programs convince large electricity users to move power use down to portions of the day where more capacity is available, such as 4:00 a.m. rather than 4:00 p.m. and thus, better distribution demand over the day.

That is economical for the customer because power can be made available much more cheaply at four o'clock in the morning than it is at four o'clock in the afternoon, and in the event that real-time pricing becomes a factor in the U.S. electricity market, it will become even more cost effective for them to not use power at peak times.

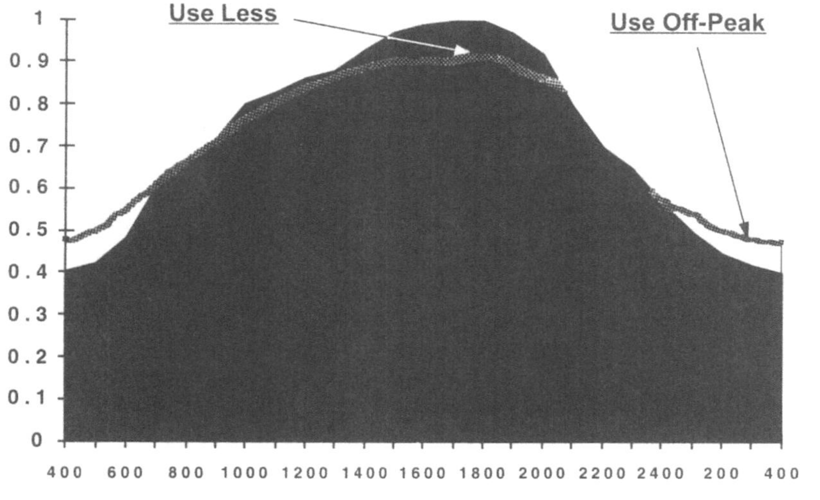

Figure 8. Reshaping the demand curve with DSM.

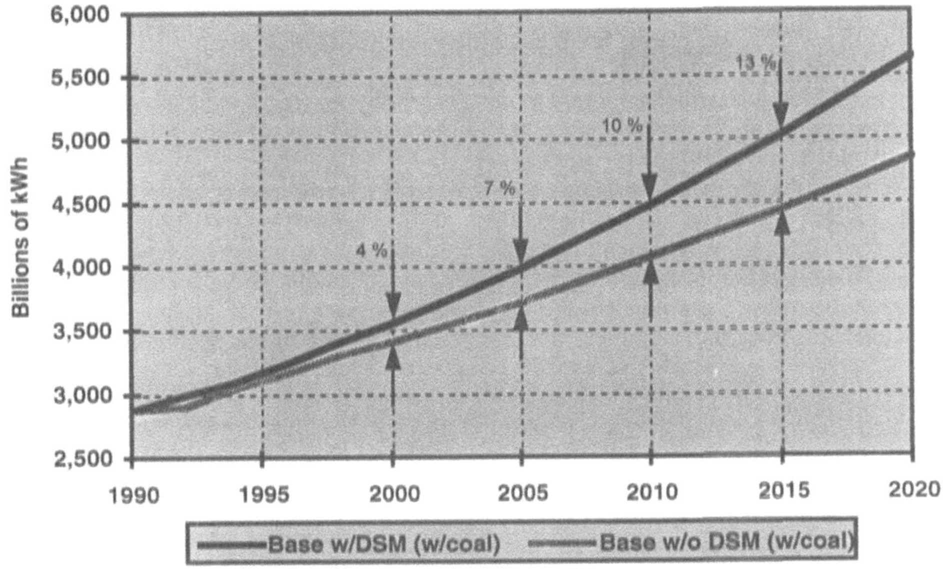

Figure 9. Net utility generation requirements. Projections beyond the year 2000 should be used primarily for trending and comparison rather than as a source of absolute values.

But the reason I brought up DSM at all, is that all the forecasts that I have presented assume a very significant savings in electric power generation through DSM activities. This is EIA data, and you can see that out in 2015, they are projecting something like 15, 16, 17 percent reduction in electrical demand just through the implementation of DSM.

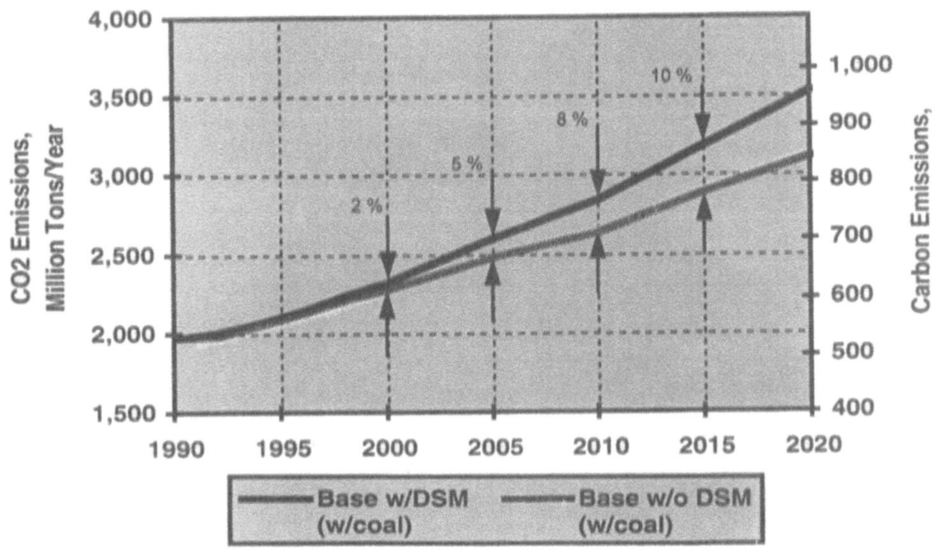

Figure 10. Summary of utility CO_2 emission rates effect of DSM. (Projections beyond the year 2000 should be used primarily for trending and comparison rather than as a source of absolute values).

Compliance Method	Number of Generators	Nameplate Capacity MW
Fuel Switching and/or blending	162	53,203
Obtaining Additional Allowances	39	14,137
Installing Flue Gas Desulfurization	27	14,101
Using Previously Implemented Controls	25	6,092
Retiring Facilities	7	1,342
Boiler Repowering	1	113
Total	251	88,989

Figure 11. What is the Clean Air Act compliance trend?

Unfortunately, that only leads to about a 10 percent reduction in CO_2 because DSM is a two-edged sword. It cuts demand, but demand is what drives the need for new capacity. New capacity additions bring higher technology, better thermal efficiency. So, while DSM does delay the onset of demand growth, it also delays the introduction and widespread use of new higher-efficiency capacity under the current scheme of operations. That is why I consider DSM to be a two-edged sword in terms of the things we are trying to accomplish.

The important thing here is that DSM in most forecasts is assumed to continue to increase over time. With the onset of competition in the electric utility industry, there is a real question in my mind whether these DSM forecasts will be met, simply because utilities in a competitive world are going to seek more customers and seek more sales and DSM activities, many of which will not be cost effective in a competitive world and will be under a great deal of pressure not to be implemented. Thus, the forecast that I showed you on that previous graph may overstate DSM impacts in that new competitive world.

So, what is going to drive change? Well, we have an old power plant fleet in this country. We have not made significant capital investments in terms of replacement capacity for quite a while. We are riding the crest of the investment wave of the 1960s and 1970s. You have heard about the gas bubble? This is the power plant bubble that we are in right now.

Figure 11 shows the history of how the electric utility industry has tried to comply with the Clean Air Act which reflects a desire not to build new capacity. What you have here are the numbers of generating units which are being moved into compliance with the Clear Air Act by different techniques. The ones at the top tend to be those which use operating cost fixes rather than capital investment fixes. Alternatives such as boiler repowering, retiring old facilities, installing flue gas desulfurization, which require capital investment, are not favored. What is favored, is fixing things by operating cost additions rather than capital investments. This strategy is just a reflection of the desire to delay new capital investments until we understand the brave new world the utility industry is going to be marching into through this decade.

Figure 12 shows you, in terms of the right-hand bars, the rate of retirement of old capacity that we would have predicted a few years ago when plants were retired after about 40 years of operating date. Many utilities are now talking about extending lifetimes to 55 years. That would delay the retirements to that set of bars on the left-hand. The real world is going to be somewhere in between. Some utilities are going to stay on the right-hand line, others are going to try for the left. The national average would be somewhere in the middle. But it is all of these new pressures, all of these uncertainties about the future which are delaying the industry's response in terms of new capacity and new technologies for many compliance actions. This effect is shown in the next slide which documents the current plans for new capacity in terms of what utilities have told the various regulatory agencies needed

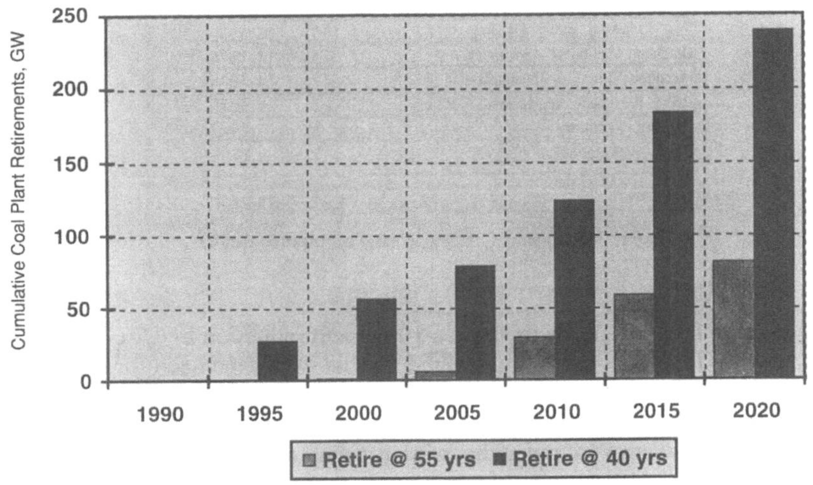

Figure 12. Coal-fired power plant retirements.

to satisfy future demands. The significant message here is that, of the identified capacity additions, about 65 percent are yet to begin construction.

Given the fact that it takes about six or seven years for base load power plants to be permitted, constructed and put into operation, you can see we are pushing the time line here considerably. And this, again, is just an indication of all the uncertainties that we have for the future.

I tried to summarize in figure 13 what is impeding the change. It is basically this: the generating portion of our industry is becoming a competitive commodity industry. Utilities today that are regulated and make capital investment decisions based on regulated least-cost economics may find that three or four years from now that new investment plant has to operate in a competitive environment, and it may not be economical in that environment.

Secondly, what we have learned in all of our environmental modeling is that there is a very strong interaction between plans to reduce sulfur, plans to reduce nitrogen, and plans to reduce CO_2. If you make an optimum cost strategy for SO_2 and five years later had to do something about NO_X, and two years later something about CO_2, you may wind up spending a great deal more money than if you followed a plan for integrated management of all three. This means that it is very important that utility management understand what the rules are

- **Generation capital investments made in a rate-based economy using least cost selection procedures may have to be recovered in a competitive commodity priced marketplace.**
- **Uncertainty regarding cost interactions between NOx, toxics, and CO_2 mitigation requirements and actions.**
- **Reluctance to add capital investment in those utilities already struggling with large capital investment burdens.**

Figure 13. What is impeding the change?

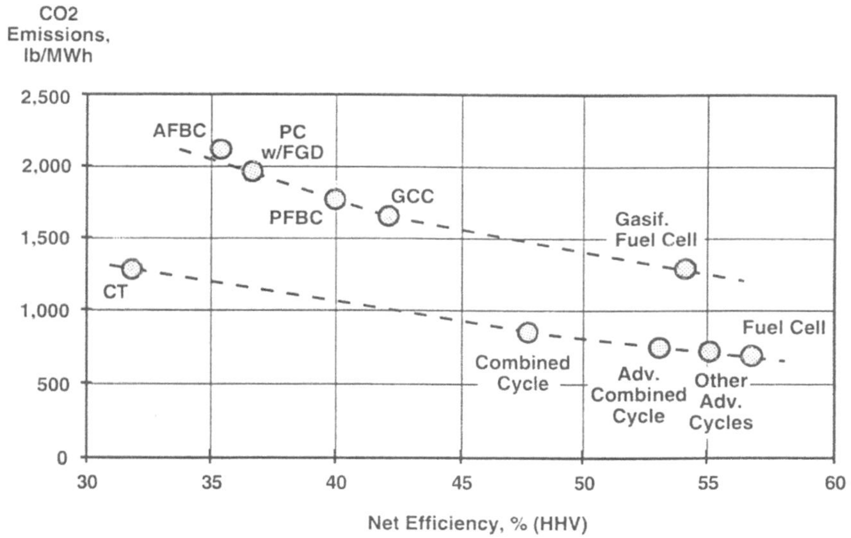

Figure 14. CO_2 emissions from current and future fossil-fueled technologies (based on average efficiencies for current and near term, and on best projected efficiencies for future technologies).

in order to make those decisions because in the end, you and I, as either the electricity consumer or the utility stockholder, will pay the bill.

The last thing, of course, is that many utilities are struggling under a very high capital cost investment burden. A significant fraction of the cost of electricity to their customers is recovering sunk capital costs. Any addition to capacity, under current economic rules, adds to that investment rate base and raises electricity rates. The only alternative a utility would have is to make that investment and then try to pay for it out of current earnings in an environment where almost every utility is being pressured to lower their rates rather than raise them; again, another indication that it is very difficult for the decision makers in our industry to respond.

Figure 14 summarizes the environmental performance of the generating technologies of the future. The top line shows the thermal efficiencies for coal-fired technologies, both current and in the future. You see that we are predicting somewhere in the year 2000-plus, that a coal-gasification plant using fuel cells would reach a thermal efficiency at close to 55 percent. We can also say that a coal gasification plant using advanced combined cycles would also reach that.

Down at the bottom you see the gas-fired technologies. The difference between those two lines, of course, is the net thermal penalty for converting coal into a gaseous fuel.

We now believe that some of the advanced combined cycles that we are presently investigating, that is, can reach thermal efficiencies of 60 percent some time early in the next century. So, the technologies will be available at that time in the future when the decisions can be made and the rules are clear for implementing them.

The gas industry has said, "Why isn't the utility industry abandoning all these coal plants; why aren't we using gas?" Well, the next slide gives you a very simple answer. This slide shows the major components of the cost of electricity for each of these technologies. The yellow bar represents the fraction of the price which is the fuel price. And you can see for the four coal technologies about a third of the cost of electricity is due to the cost of coal. For the gas technology, fully two-thirds of the price is due to the fuel price.

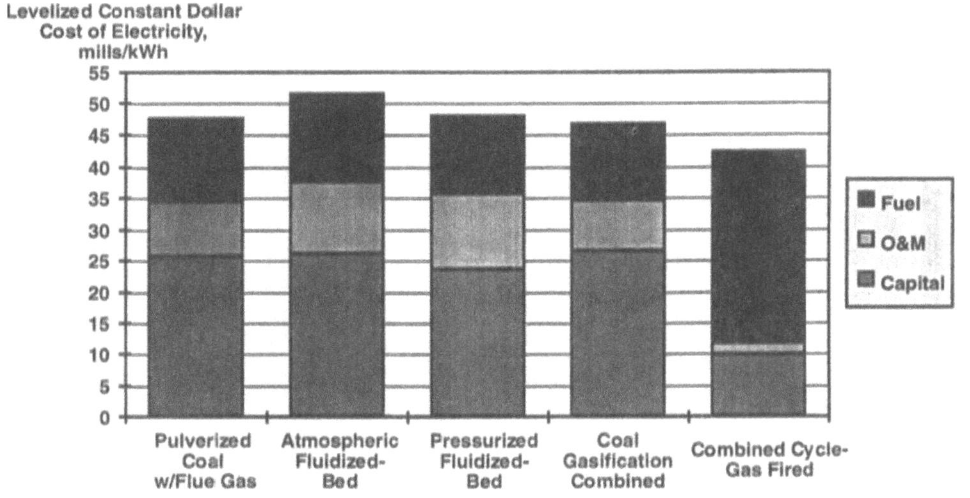

Figure 15. Cost of electricity from near-term fossil power technologies (based on bituminous coal with 4.0% sulfur). [Based on average C.O.E. parameters for each technology.]

Now, I am not going to try to forecast what will happen to the price of natural gas if utilities double consumption, but I suspect it might go up. And it is that risk that keeps utilities from running headlong into gas fuel technologies. Traditionally, we have held the line on electricity prices by having diversity in the fuel mix. When gas prices go up, we shift to coal or use nuclear more or use hydro more. If you become overly dependent on one fuel, you become captive to the price fluctuations in that fuel. We will not look very good to you if we do that.

This is true even for the new technologies, the ones that we hope will come on line in the next 10 to 15 years. All of the new gas-fired technologies, while they have very, very high thermal efficiencies, retain this characteristic where the price of electricity is extremely dependent on the price of gas.

Therefore, our argument is that a balanced portfolio which includes coal, and if CO_2 becomes a problem, probably must include nuclear, we are truly going to maintain a cap on emissions and a cap on price, and continue reliability and flexibility.

What will drive the change? Well, for those utilities that don't de-regulate quickly, for those that sit back and say, we are going to watch those crazy Californians to see what happens to the price of electricity there, these are really the things that are going to drive things—not everything, I can't be all-inclusive.

I believe that we must continue R&D, particularly for the non-emitters. We have to find a way to use effectively non-emitting technologies that are cost effective. We are not against renewables in our industry. At EPRI, we have spent a tremendous amount of money trying to make solar cost effective. We have not succeeded, not as a base load or large-scale technology. There are places in the world, places in this country where solar makes a great deal of sense, but not generally.

The same thing is true of wind. In a lot of places it makes a great deal of sense, but not generally.

We haven't solved the problem of storage. If we could devise some great universal container for electricity so that we could build three times as much solar and store the

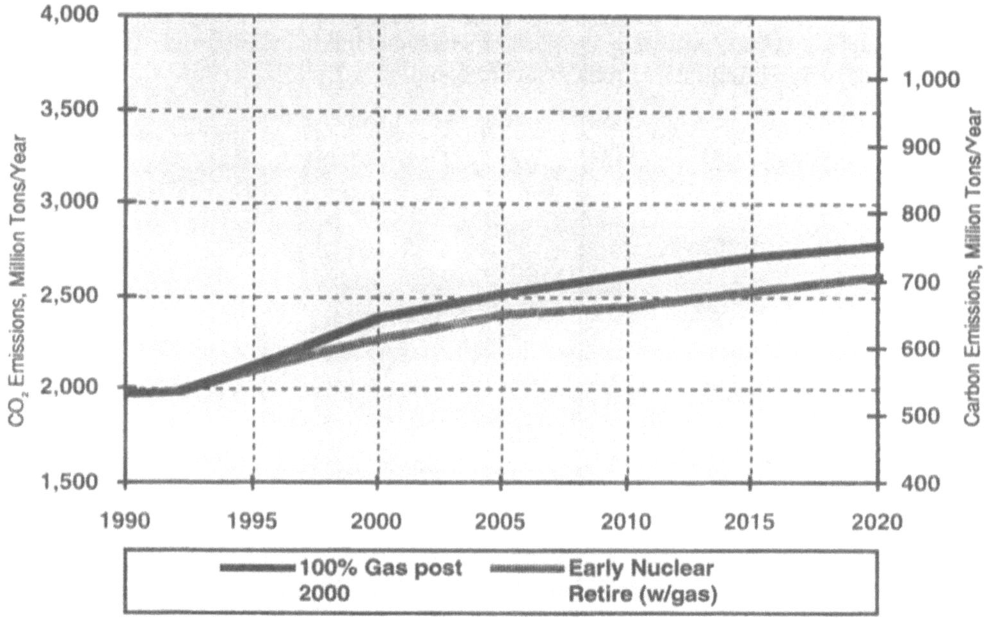

Figure 16. Summary of utility CO_2 emission rates effect of early nuclear retirements. (Projections beyond the year 2000 should be used primarily for trending and comparison rather than as a source of absolute values.

electricity for when we don't have sun, everything would be wonderful. But we haven't solved that problem. It is a key problem for the future.

In an unregulated utility economy things become even more confused. Again, deregulation is a two-edged sword. It will give utilities more flexibility to do least cost types of things, recapture capital investments over much shorter periods of time. But at the same time, the emphasis will be on the least costs, it will be on lower capital investments which don't make renewables right now look very good. It will be on recovering those investments over a shorter period of time, which means that those people will want to run base load, they will want to maximize their output, which gets back to whether DSM will continue to be something pushed by competitive utilities. I can't predict the answer. I am just giving you the conditions.

Let me wrap up with this final graph because it again illustrates the vulnerability of many of the forecasts that people make. The green line is CO_2 emissions if we keep all our nuclear plants in operation until their scheduled retirement date.

In a competitive world, it will be much more difficult for utilities that own only one nuclear plant to do that, because the cost burden for all of the overhead costs such as relicensing, safety and for a utility that owns one nuclear plant, is about the same as for a utility that owns three or four. Clearly, the cost of meeting regulatory requirements is much higher per nuclear kilowatt hour than for the utility which owns multiple nuclear plants.

As utilities face decisions about major investments in those nuclear plants to keep them running, it becomes problematical as to whether the decision will be yes.

This slide shows you what would happen if all utilities shut down their nuclear plants at an earlier retirement date than when they are scheduled. And you can see that the net result in CO_2 emissions is to raise them significantly.

So, as we sit and think about the future and look at all of these forecasts, we have to keep in mind that they are only people's guesses as to what might happen, and generally they exclude the synergies and interactions in a very complex world.

17

SESSION III DISCUSSION WITH CHARLES SIEBENTHAL

Mr. Fox-Penner: You showed a slide from the North American Electricity Reliability Council. By my quick glance, there was about 300 million metric tons worth of coal capacity added between 2000 and the end of your chart.

Mr. Siebenthal: That's CO_2 emissions.

Mr. Fox-Penner: Right, CO_2, which says that there is another 300 million metric tons worth of coal on the drawing boards today.

Mr. Siebenthal: That's correct.

Mr. Fox-Penner: And you so no reason why that is changing or no reason to question those data.

Mr. Siebenthal: I can't question those data, because they're not mine. You have to understand that those data represent commitments made by utilities to their Regulatory Commissions about the need for future capacity and what they might build if they had to build that capacity today. There is no commitment made by those utilities that what they install will be that particular technology.

Clearly, if the decision to build some of that capacity is made five years from now and natural gas prices are still where they are, a lot of that coal capacity will come on line as gas which would change the numbers significantly.

Mr. Fox-Penner: And do you think restructuring could impact that?

Mr. Siebenthal: I would say restructuring will definitely impact it, but I can't predict how. The entire decision-making process that utilities go through is changing. As we delay a capacity, all we're doing is saying that instead of building power plants at kind of a steady rate so that the prices remain constant and so that the infrastructure for fuel supply can be added to incrementally, we're going to need to do a lot all of a sudden. It's the normal U.S. economic problem of not having a planned rate of attack, the way they do in Europe.

If we decide to quadruple the rate at which we build power plants in the year 2000, the net impact will be that there is going to be a shortage of people to do that. There's going to be a shortage of regulators to pass the regulations, and it's going to slow things down. But the real impact is that the price is going to go up. And how the individual people who are generating electricity five years from now, whoever they are, are going to make those decisions and respond to that environment, I cannot predict.

Mr. Fox-Penner: You also showed somewhere one to two hundred million metric tons worth of demand side management, if I read it right.

Mr. Siebenthal: Yes.

Mr. Fox-Penner: And you said restructuring may affect that amount. Was that based on someone's estimate of what is cost effective to do, looking out into the future as things stand now? In other words, can we assume that that was cost-effective DSM in the economic interest of the participants?

Mr. Siebenthal: No. That was regulatory DSM, an estimate based on the implementation of DSM as it's been done for the last five or six years, where regulatory commissions require utilities to make investments to reduce demand and put that in the rate base, whether or not there's a confirmation that that is truly cost effective either to the consumer or to the utility. It's based on a hope that it's cost effective, and in many cases it has proven to be cost effective. But there's no guarantee of it.

I raised the question about DSM because a few of the utilities who are very eager to get into retail competition, low-cost providers and who are run by executives who used to work in the natural gas industry, have filed DSM programs with their regulatory commissions, which are significantly lower than what the regulators would like to see. They are looking at their cost to provide electricity and trying to cut out anything that doesn't have demonstrated value in return. And some of the DSM activities that have been purely regulatory driven have been classified in that category by those utilities.

I can only imagine that as more utilities get into cost cutting and into that competitive mode of thinking, they will raise the same questions about many of the DSM activities. I can't predict what they will do. I'm just raising the issue that you cannot rely on DSM as much and with the same degree of certainty that we would have in a regulated world.

18

INDUSTRIAL DEMAND

The Climate Wise Program

Pam Herman

U.S. Environmental Protection Agency
Climate Wise Program, OPPE
401 M St. SW
Washington, DC 20460

We're finding in the industrial sector, which accounts for about a third of U.S. energy use, some amazing cost savings from energy efficiency; a real opportunity, through energy efficiency, renewable energy and innovation that leads to operational efficiencies, real cost savings, and real benefit to the bottom line. In many cases a program is keeping facilities up and running, creating jobs where facilities were previously scheduled for closure because they were operating so inefficiently.

I would challenge the thinking that says we can't afford to retire existing ways of doing business or existing old capacity for new ways of doing business.

Efficiency and the need for improved efficiency is really driving U.S. industry in the direction of innovation, although I will say that many of the reductions that we are finding are really low tech solutions.

Let me talk about Climate Wise to give you the context that I am working in. You heard from Karl Hausker and Dave Rinebolt yesterday about the Climate Change Action Plan that was launched by the President in response to the 1992 Rio Convention.

The Climate Change Action Plan was released in October of 1993, and the Climate Wise Program is one of the foundation actions. It's a broad-based program like Climate Challenge. It provides an opening across sectors for organizations to participate in the Plan, whose goal is to return emissions to 1990 levels by the year 2000.

Actions that industries are looking at under the Climate Change Action Plan include some of the following programs. The Motor Challenge Program, which Dave talked about yesterday,is a real success story in signing up companies to replace existing motor systems with more energy efficient motor systems. And they're also looking at market pull opportunities for encouraging the development of new, more efficient motor system technologies. The Green Lights Program, which is EPA's flagship Voluntary Energy Efficiency Pollution Prevention Program, has been enormously successful, and over 1600 companies are working through the Green Lights Program to install energy efficient lighting in their commercial office space, as well as some of their industrial space.

Evaluating Climate Change Action Plans, edited by White
Plenum Press, New York, 1996

The Waste Wise Program is, like Climate Wise, one of EPA's newest programs. It encourages recycling, source reduction and the purchase of recycled materials to reduce both CO_2 emissions in the development of new virgin materials, and also to reduce methane emissions from landfills by reducing landfill source reduction.

The Energy Star Buildings Program and Rebuild America target residential and commercial building structure, for energy efficiency renovations. Right now, Energy Star Buildings is working with companies at 25 demonstration sites that are demonstrating 25 percent efficiency improvements from these retrofits.

The Energy Assessment and Diagnostic Centers (EADC) are primarily targeted for medium and small industries. These centers are run through universities around the country, I think 30 universities. The goal is to expand those to about 100. EADC Centers provide on-site auditing assistance to small and medium-sized firms, and they've been really very successful. There's a wealth of information from the EADC audits on industrial and manufacturing efficiency opportunities. It's a really successful program with 65% of EADC audit recommendations are accepted and implemented by the participating company.

In addition, the Center's are training energy managers of the future by involving graduate students in energy efficiency auditing. So we are really training the future energy managers and environmental managers. A number of the EADCs have expanded from a focus exclusively on energy to environmental audits, as well.

And then Nice^3 is a financial demonstration, a grant program, that states' energy and environmental offices run cooperatively to demonstrate pollution prevention and cost-saving technologies.

Climate Wise focuses on these large industries that are putting together a comprehensive portfolio of emissions reduction actions and helps them to select the action plan initiatives that are most appropriate for their organization and encourages them to go beyond those initiatives to innovate outside of the action plan.

Then we share the results of their innovation across the country. We provide public recognition, the technical assistance to help grease the skids, if you will, and our goal is to work with about 650 companies and capture about 20 percent of U.S. industrial energy use. Right now, in the eight months since we've been in existence, we've got about three percent of U.S. industrial energy use. So we're doing pretty well.

Who's involved? It really is a true partnership between EPA and DOE. We work together on a daily basis, and the synergism between the two agencies allows us to bring together DOE's great research and informational resources with EPA's environmental perspective and fast deployment capability, providing a great vehicle for launching this program.

In addition to working with the large industries, we're working with trade associations, such as Business for Social Responsibility and the Chemical Manufacturers Association and others to try to get the word out about the program, cost -effectively and efficiently, for our industrial customers.

We're also working with cities and states. We're developing a pilot program for a state deployment arm. We've heard Dave Rinebolt's presentation earlier and know that the states provide a wonderful deployment capability that the federal government obviously just can't match. We are looking at a pilot program in which we'll work with 7 states to provide the on-site technical assistance that industries will need and are asking for to launch these programs. And we think that that will be very successful, given the state capacity.

We're also working to build bridges between the State Energy Offices, the State Environmental Offices, and Economic Development Agencies, and requiring that as part of participation in this pilot program to again build more synergies between energy and environment.

Our cities initiative is trying to get cities involved in working with their industrial partners to improve efficiency, prevent pollution, which is a win-win situation for both, both from an economic capacity, but also from a pollution capacity. And I'll give you an example in a little bit that talks about some of the benefits to municipalities; for example, efficiency in reducing waste water and reducing water use.

We're also working with the financial institutions. We are working cooperatively now with the Small Business Administration in Colorado. And they've put up about half a million dollars to provide 50 small businesses with loan guarantees to do energy efficiency improvements, which is really a new step for them. We hope that the work with SBA and other financial institutions will help leverage private sector funding for some of the more capital intensive efficiency opportunities that industries are facing.

So in addition to leaning on the utilities for capitalization through DSM programs, we're also trying to build a stronger financial base in the private sector.

Some of the emission reduction actions that we are targeting, in addition to participation in the sector-specific or technology-specific reduction opportunities targeted by other CCAP programs include, process improvements such as fuel switching, use of renewables, employing mass transit or carpooling programs, carbon sequestration, advanced technologies and co-generation. Again, it's really sort of encapsulating a lot of different opportunities that industries can bring to the table for CO_2 emission reductions.

Many of us believe that after the mid-70s, we picked all the low hanging fruit available in the energy efficiencies opportunities. But what industries are finding is that some of that fruit has grown back since the high efficiency days as fuel prices have dropped, or there's new efficiency opportunities that they're finding year in and year out. Some of the total quality management philosophies that industries and large companies have picked up is actually spurring more ideas and opportunities for innovation.

A good example of the TQM approach, once that we used in the design of the Climate Wise program, is the Dow Louisiana Division's Energy WRAP Program. This is a program that they started to involve employees in identifying energy efficiency opportunities and, over the past twelve years, they have continued to find enormous cost-saving, cost-effective opportunities for emission reductions. They continue to every year find projects with million dollar savings, with average return on investment of over 200 percent and even in '93, their cost savings have been enormous. They're looking at something like 140 projects in '93 and still meeting the return on investments that they've historically achieved over the years. That's a really great success story.

In terms of a dollar savings, the Breyers Ice Cream example is one where DSM really came to the fore. Boston Edison and Breyers Ice Cream teamed up to do a DSM program that actually saved this plant in Framingham from closure. It was about to go under, taking with it an employee base of about 150 employees. All those people were about to lose their jobs. They got together with Boston Edison, and they redesigned the plant. It was capital intensive, $3.5 million, to bring in new refrigeration systems and upgrade the plant. Now they're one of the best operations in the chain, and they've hired another 30 people as a result of their improved efficiency ;annual cost savings $425,000. They've actually recently been bought out, but they're still achieving enormous annual savings from their efficiency program.

Very Fine Juices is an example that I alluded to earlier, where the company teamed with their municipality to look for opportunities for improving their water use efficiency. They were one of the biggest users of water, and the municipality was having a hard time keeping up with the pumping and treating costs to provide water, and to deal with the waste water on the other end. So the city got together with Very Fine Juices. They came up with a recycling and waste water reduction program approaching 90% efficiency. This is the cost savings of $445,000 annually to Very Fine Juice. That does not include savings for the municipality in terms of water and waste water.

A lot of opportunities for partnerships exist. Greenfield Tap and Die was another company threatened with closure because of high energy and environmental costs but working with their utility they identified some equipment and process changes. The result, $90,000 annual energy savings, and from their hazardous waste disposal, another $180,000 of savings along with production time savings of 11 days to 1 day.

Dupont, the premier signatory to the Climate Wise Program, and recently launched the Jump Start Program to capture quick turn around energy efficiency savings.

In 31 Dupont facilities across the country, they have switched fuels, optimized performance, discarded obsolete equipment and, with almost no capital investments, have achieved $12,000,000 in savings in a 4 month period. They're estimating that annually it will reach $31 million for Dupont as a whole, which is amazing.

So back to Climate Wise, we're working with companies to get them to pledge to undertake cost effective emission reduction actions. We work with them to identify and commit to these actions and we realize, especially from the Dow example and from Dupont, that it really takes organizational support to realize these savings and that infrastructure is important.

So we're also working with companies to establish energy management systems and encouraging them to set performance improvement targets. The reason that we're not requiring goal setting is because we have found in some of our past experience with the voluntary programs that companies are going to aim low, because they don't want to lose. They don't want to overestimate and then not achieve what they said they were going to and look bad.

But if you look at the total quality management philosophy, it says if you aim high, you're likely to achieve much more than if you just try to meet the lowest level that you can. So we're trying to get companies to change the way they do business internally; to set stretch goals and really work to meet those achievements or come as close as they can. This way, they can shoot for 80 and come up with 70 and still feel good about it.

Climate Wise companies now represent about three percent of U.S. industrial energy use. Fetzer Vineyards is one of the newest companies to sign up. They completed a new plant a couple of years ago. They're using old bluejean material to insulate the plant and they constructed the plant from old buildings that were deconstructed in other parts of the country and shipped to their site to rebuild. They have pledged, among other actions, to make their operations 100% powered by renewable energy supplies by 2010.

Dupont has pledged to reduce emissions by 18 million metric tons of carbon dioxide equivalent. That's an amazing emission commitment, but it does involve some actions that we previously estimated in the plan, including HFCs. Nitrous oxide emission reductions are significantly lowered as well.

Johnson and Johnson participate in a variety of different programs and are using the Climate Wise Program as something of an umbrella. They will also be aggregating their emission reductions from participation in all of these programs through the Climate Wise Program and will have an opportunity to see where their investments are being most profitable. They have pledged a 10-percent energy reduction by 1996, 50-percent reduction in non-hazardous waste and 10-percent reduction in hazardous waste, a significant commitment.

AT&T is just completing their five-year energy management plans in all of their operations. They are looking for cost savings from innovations, including the use of landfill gas as an energy supply in one of their facilities, and are participating in some of the other voluntary programs. And they'll be estimating the actual reductions from these five-year plans. They have about 75 percent of the plans completed.

Georgia Pacific. Their pledge includes transportation options such as back hauling; container board; trading; carbon sequestration actions; and efficiency in boiler controls and operations.

Climate Wise recognizes corporate actions to protect the environment and provides technical assistance. We're working on pulling together information on technical assistance resources, both in the public sector and in the private sector. There's a lot of technical assistance aids out there, but when a company sits down to try to put together their pledge, we find that they're basically working on their own, and may not know who to call.

To facilitate this we're launching ten regional workshops, which will be held across the country starting in November to bring together companies in different regions to share success stories and to hopefully get more companies participating in these types of actions.

We are encouraging participation in Section 1605B reporting system as a vehicle for reporting emission results, rather than setting up a new system, and we're working with DOE in the design of that program under the Energy Policy Act. That will come on line in June 1995 and will begin to provide recognition for actual results.

Regarding the recognition program: Some of the recognition categories we're looking at include greatest emission reduction achievements, reductions on a sector or industry basis, historical achievements recognizing companies like Dow that have lead the way.

Let me close here with a quote from one of the companies that we're working with. Johnson and Johnson says: "Failure to participate in the voluntary programs is a vote for regulation and taxes."

ELECTRIC UTILITIES

The Climate Challenge Plan

Dirk Forrister

United States Department of Energy
1000 Independence Ave., SW
Washington, DC 20585

Niagara Mohawk, a utility from upper New York State along the Canadian border, and Arizona Public Service Company, an electric utility along the Mexican border have made an agreement with each other.

And what made these companies do this? What made a company from Arizona transfer 25,000 perfectly good SO_2 allowances with a market value of \$140 to \$150 a ton, total value of roughly \$3 million, to a utility in upper New York State, who would transfer back to them in exchange 1.75 million metric tons of carbon reduction? Why would they do that? And why would the utility in New York then take this perfectly good asset with that market value and donate it to a public interest group, where it could get a tax deduction. They would then apply those tax reduction proceeds, which will be roughly a million dollars, to further efficiency improvements in their own service territory and in some joint implementation pilot projects, which they would hope to submit to the U.S. initiative on joint implementation.

Are they crazy? Why would they do that in this marketplace? I'm going to try to give you some illumination of it, because the reasons are that both of these companies have made commitments to the United States government to improve on their greenhouse gas emissions. They are utilities that are situated differently in that regard. One of them is in a position to be an over-achiever, and the other one is in a growth region that's really going to have trouble meeting a stabilization commitment which it made to the Secretary of Energy. Arizona is in a growth area.

Their deal was to come up with a backstop for Arizona's plan, to have some tons in a bank account, if you will, so that if they have trouble in any given year they can draw on that bank account. And in some sense, it's a form of cooperation that was implicit in this Climate Challenge Program, but which was not planned for or scheduled. It's really a form of domestic joint implementation, where they can cooperate and do more together than either would be able to do alone.

Where this began really is with the Secretary of Energy and the interagency team that developed the President's Climate Change Action Plan, that believed that voluntary was an appropriate way to go. The Secretary of Energy's over-arching philosophy made room for

Evaluating Climate Change Action Plans, edited by White
Plenum Press, New York, 1996

this. In this new phase of environmental protection, from her place in industry and her place previously in government, she had regulated prices of certain fuels in this country. She believed the better way to go in a regulatory sense was for government just to announce a goal of what needed to be done and to ask industry to help meet it – to ask industry to enter written commitments of what they're willing to do to help us get there, but allow them flexibility.

That's the over-arching policy goal that's behind the Climate Wise Program and that's behind the Climate Challenge Program. These things are twin programs. The Climate Challenge Program has been out the gates a little longer and has a little higher level of participation. We've got about 80 percent of the electric utilities in the United States participating, and the electric utility emissions of greenhouse gases in the U.S. are about 36 percent of our total. It's the largest single emission sector.

How many of you have heard a presentation on the Climate Challenge before? How many of you participated in the development and implementation of the Climate Challenge Program? Some of you know the program probably as well or better than I do, because you've been doing it inside your companies. I hope you know who each other are, because you share are a wealth of information.

The President announced this program and the original action plan in October of 1993. By that date we had, I think, just over 600 companies signed up for the program and by Earth Day of this year, we had right around 800 utilities.

Between the original signing of the letters and submittal of letters to the Secretary of Energy from CEOs of electric utilities who were submitting letters of interest to participate in the program, we developed an umbrella agreement between the largest trade associations in the United States and the federal electric utility, the Tennessee Valley Authority.

We announced an agreement among five players in the industry and the Secretary of Energy on Earth Day of this year in the midst of an enormous technology display on the Capitol Mall in Washington. We announced essentially what our game plan was, which set forth basic parameters for the program- what we were going to seek to do. The Vice-President attended and actually signed the agreement himself in a moment of excitement.

It basically pledged the Department of Energy to be an honest record keeper for utilities who were making voluntary commitments and to do so under a program with the Environmental Protection Agency called the 1605B Program of the Energy Policy Act.

Our job was to get out the regulations, the accounting regulations, if you will, for how a voluntary emission reduction database would be set up, how we would try to score reductions in the Energy Information Administration, which is the basic energy record keeper of the United States government.

We kept our part of that bargain in that these guidelines have now been promulgated. A whole series of hearings were held across the country, and these are publicly available on the Internet and you can also get them from the Department of Energy's Policy Office.

We have also provided guidance to the Energy Information Administration on how to set up the database. They have already promulgated reporting packages, forms, for people to report their voluntary emissions. And there are two versions of these forms.

There's one for Boy and Girl Scout troops, if you will, or small businesses, called the Form EIA 1605EZ reporting form. It is for small reporters that have simple things to report, not too complicated, and there's a set of guidelines.

We also have a more thorough form, which is the one that I expect most of the larger Climate Wise/Climate Challenge type reporters would use, the basic 1605 reporting form. These are out in draft form right now and will be final in the spring of 1995, because we're hoping to start logging reductions in the April time frame of next year. We're currently in the process of getting forms approved and taking public input on forms.

Those are some of the things that we in the Department of Energy have done. Most of the work, though, has not been done inside the Department of Energy. Most of the work on this program has been done at the company level. Companies like Illinois Power have put together their own teams to look across their systems to see what are cost-effective things that they could do within their systems to improve on greenhouse gases and make a contribution to this database and to the voluntary efforts in the United States.

Porter Womeldorff, of Illinois Power, led a group with some of my colleagues at the Department of Energy, coming up with a menu of possibilities of things that companies ought to consider. We call it the Options Workbook, and it is publicly available now from the Department of Energy. It's got a whole host of ideas for electric utilities to consider – things that they may be able to do working in partnership with their customers on demand-side measures or on their own transmission and distribution systems, ways to improve efficiencies, supply-side improvements or moves toward renewable energy and methane capture. Some of our electric utilities in the United States have their own coal mining operations, or some of them are municipally owned and have their own landfill operations, where there are methane capture possibilities.

There are other things electric utilities have in this country. Many of them have vast land holdings and rights of way that are huge, so there are opportunities for reforestation, tree planting efforts, and other kinds of biomass programs.

The Options Workbook has been put together. We had agreed that after the Department of Energy put out its 1605B guidelines for reporting, beginning 60 days after that, we would be starting agreements. I'm frustrated that I can't tell you what's in any of these agreements because, at this writing, the 60 day clock is still ticking. Soon we will start seeing plans from individual companies.

We have a number of electric utilities that are in active talks with the Department of Energy about what's going to go in their agreement. Companies are coming in every week to sit down with a team of DOE experts to present what's going to be in their company's specific action plan and to get our ideas about things that might be clarified or improved. They are in the process of planning for a public announcement of these things.

As a person who has been involved with this program for a real long time, I was hoping that the utilities could come in and deliver as promised with creative and innovative solutions that we in government didn't think about, or maybe even the people from the industry who put together the Options Workbook didn't think about.

True to form, we are seeing a whole host of new portfolio approaches. We are seeing things that we in government or representatives of companies who come to meetings in Washington didn't dream of, but things that operational people who work in the field every day, did come up with that are going to contribute. Maybe they aren't ton contributors but they are pound contributors, and, as you know, we'll take anything we can get in this game.

We are seeing creativity and innovation exhibited. And as a matter of fact, there's even a little competition between these companies, believe it or not, to be different from one another and have something in their plan that nobody else thought about. My hope is that once those things become public, companies consider taking on additional actions that their friends in other service territories are adopting.

Some of these plans are more ambitious than you might have expected. Others are less ambitious than you'd want. But the common theme of all of them is that they are progress. They are bringing new tons to the table that, without this program, we would not have achieved. No matter how many technology-specific programs we tried to fund to stimulate activity, you wouldn't have seen some of these tons come in. They're really innovative and exciting things.

This plan is good for the public, because it is delivering cheap tons that are going to keep rate payer's bills low. This is a plan that is going to deliver more tons. In the changed political atmosphere, I've been heartened by not seeing any change in the way companies are approaching this. It fits with the quality ethic that companies are striving for – continuous improvement in their operation. The electric utilities have been on this program long enough to get committed to establishing the framework for this type of approach, not just on this program in the climate change area, but they would like to demonstrate its validity in other areas. They recognize that it's a golden opportunity to make a voluntary system work, and that if it doesn't work, it may be a long time before they get that chance again.

There is no backing off, because they would like to prove out this framework for the future, because they'd like to avoid the intrusiveness of regulatory programs that may clamp down on them from the federal government or at the state level.

Finally, it's a good deal for the government as well, because we aren't having to figure out what those regulatory things are that we ought to be clamping down on, and we are getting the results that we want in a much quicker time frame.

The Secretary frequently says that if we had taken the approach of going to Capitol Hill for legislation in the last session of Congress, it would have taken five years from proposing legislation, getting it through the Congress to getting the regs written and implemented. We are still seeing pieces of the Clean Air Act get fully implemented, four years after the fact.

We're getting much faster results out of this type of approach than we would have if we had gone the legislative route and we think it's got validity, not only on this problem of climate change, but perhaps in other areas. So we're extremely enthusiastic about it.

20

DISCUSSION WITH DIRK FORRISTER AND PAMELA HERMAN

Mr. Fox-Penner: "Are there any proposals you are aware of for bringing back energy credits for federal income tax purposes?"

Mr. Forrister: Probably, it's too early at this point to know. I've heard questions asked about whether it's time, but I know of no specific credit proposal that's being floated by any member of Congress. They don't even know what committee it would go to if it were proposed, I guess it's Ways and Means, but you never know.

Mr. Fox-Penner: A question for Pam Herman: How do the Climate Wise and Climate Challenge Programs coordinate? Will utilities participate in both programs, or is it an either/or situation?

Ms. Herman: Utilities are participating in Climate Challenge. We're not signing up utilities in the Climate Wise Program. Individual industries participate and individual companies participate in Climate Wise. We're trying to work together with utilities, providing them with information on Climate Wise, and they are using that information to market their DSM programs and their outreach programs with their customers, large industries. They are encouraging participation in these programs by offering them participation in the DSM and an opportunity for national recognition under the Climate Wise Programs.

We are really trying to build partnerships between large industries, between industries and utilities, but the programs are separate in terms of their outreach.We're also taking great advantage of the Options Workbook that Climate Challenge put together to help give some ideas to industrial customers as well.

Mr. Fox-Penner: Do capital cost estimates for nuclear power plants include costs of decommissioning old plants and waste disposal?

Mr. Siebenthal: They try to. I don't think anyone really knows what a true cost for decommissioning a nuclear power plant is today. We have a project under way at EPRI to try to provide what we hope will be a realistic estimate. But since the regulations in any state for final disposal of everything really don't exist, you're really shooting in the wind in terms of trying to figure out what the costs will be.

We do provide a reserve in those costs. Whether or not it will be adequate to eventually achieve a complete decontamination of a site is anyone's guess. I would suspect not, but that is the best we can do at the present time.

Mr. Fox-Penner: Regarding Clean Air Act amendments, as utilities reduce SO_2 and NOX emissions, do they also achieve CO_2 reductions as a byproduct? And can't utilities take credit in the future for these CO_2 cuts? How are greenhouse gas emission reductions associated with compliance with the Clean Air Act treated with respect to voluntary greenhouse gas reduction actions?

Mr. Forrister: The Climate Challenge Program for utilities is built to feed tons into the 1605B database that I mentioned. Let me tell you where that program came from. It came from an interest on Capitol Hill — I know because I was there at the time — sending a signal to electric utilities to be considering climate change impacts of their Clean Air Act decisions.

There's nothing in the Clean Air Act to directly require companies, in their acid rain compliance, for example, to adopt strategies that help with greenhouse gas emissions. As a matter of fact, there are options there that worsen their greenhouse gases. Members of Congress didn't want to tamper with that and say not to use, for example, a scrubber technology that might have an energy penalty associated with it that's going to result in an overall increase in emissions. They wanted to provide a signal that they would like companies to look for things that would also improve greenhouse gases.

There are options in the Options Workbook for boiler improvement that can, for example, take the sulphur out earlier in the process so that there is an energy efficiency gain and a CO_2 gain.

There are certainly benefits of demand-side measures that can do the same thing. They can result in overall reduction in an old, inefficient coal plant, for example, so that both your SO_2, NOX and CO_2 emissions decrease.

The only thing in the Clean Air Act that was also a signal to utilities to be thinking about was a requirement for companies, as they monitor for their SO_2, to also report the monitoring results of their CO_2. There's a signal there, but no explicit requirement. Usually the same people that are trying to come up with a Climate Challenge portfolio are involved in the decision making about what their compliance strategies are for the Clean Air Act. That fits a theme that EPA is pushing in some other areas on the common sense initiative, to look at broader problems than just the Clean Air Act and climate change.

Ms. Herman: EPACT and the 1605B guidelines, are being designed to be flexible so that the large utilities that are preparing Clean Air Act reports, can also submit those for recognition under EPACT as well; trying not to be duplicative there, but to marry the two.

What Climate Challenge does on the voluntary side is to work with companies to help them achieve reductions.

Mr. Siebenthal: I think this issue of the synergy between reductions and emissions is one of the reasons again why utilities have delayed their decisions. Several of the technologies developed for application under the Clean Air Act for SO_X and NO_X controls, such as pressurized, fluidized bed combustion and atmospheric fluidized bed combustion, utilize limestone as a way to capture sulphur. Those processes have higher CO_2 emissions than a pulverized coal-fired power plant with a scrubber. They're cheap. They have nice efficiencies. They work very well, but if CO_2 limits become a problem, no utility would want to buy those technologies and implement them in the event that they were going to bump up against a CO_2 emissions requirement.

There are also technologies developed for retrofit applications, which is another reason why you haven't seen very many retrofits. The continual evolution of more regulation on emissions always makes the decision-making process more complex.

Mr. Fox-Penner: The pledges under Climate Wise, i.e., the 20 million metric tons by Dupont, first of all, is it CO_2 or carbon equivalent? Are they cumulative or annual? And what is the baseline?

Ms. Herman: Most of the companies are looking at CO_2 equivalent and not carbon. The Dupont pledge, however, is carbon equivalent, but it's varying, depending on the company's pledge. And the 20 million is cumulative. It's not annual. The base line is 1990.

Mr. Fox-Penner: At a conference a few weeks back, Sam Skinner, President of, Pacific Gas and Electric stated, "DSM is only a short-term remedy, hardly a long-term solution." Do you agree with this, and what about the side benefits of DSM, such as decreased dependency on foreign oil, economic development benefits, increased comfort and convenience, etc.?

Mr. Siebenthal: Whenever you talk about DSM, it covers such a range of activities that you have to quantify what you're talking about, and I don't know what Mr. Skinner was trying to speak about. DSM includes things that Ms. Herman discussed, physically putting devices out in a customer's plant or home to reduce consumption of electricity. With such a device in a plant or in your home the utility can press a button and limit your use of electricity. You may rebel and decide that you want to pay $10 a kilowatt hour to keep your air conditioner up at 4:00 in the afternoon, when it's 120 degrees. We may give you that opportunity sometime in the future.

Clearly, direct DSM has long-term effects. Shorter term activities, one can only guess about. Again, they cover such a multitude of things, I find it difficult to quantify an answer to the question.

Mr. Fox-Penner: Did Secretary O'Leary tell the White House Conference on Climate Change that her preferred technological mitigation option was carbon expansion, and what percentage of JI proposals do you think will be carbon sink projects?

Mr. Forrister: The Secretary is very much committed to portfolio approaches, because she knows that there are instances where any single type of a program can under-achieve, she'd like to have a lot of options out there that can over-achieve as well.

What percentage of the JI submissions would ultimately be sinks? I have no basis to judge as yet, because, although proposals are in and staff are working with submitters on fleshing out information, I don't expect it to be a dominate majority. I expect it to be part of the portfolio. I should make clear that Climate Challenge participants are permitted to include Joint Implementation pilot projects as part of their commitment, whatever that commitment might be. And so far, I'm not seeing a whole lot of international tree projects in the submissions coming in with company-specific plans. Nor am I seeing a lot of international projects altogether.

They're interested in doing things in their own service territory first, and they are also interested to see how the international community resolves some of the questions before they invest precious money in this increasingly competitive atmosphere. They are looking at places where they are getting strong signals of interest from the host governments.

Mr. Fox-Penner: What about the risk associated with nuclear power, and the near total lack of fuel price risk with solar and wind? What about the cost of operation and, to a growing degree, reliability of these energy sources?

Mr. Siebenthal: That's a multifaceted question, and I'm not sure I understand it all. I did not attempt to deal with all of the risks, and those risks certainly need to be grounded to anybody's decision-making process. I attempted to illustrate only a single risk about over-dependence on gas.

One has to ask what risk means for solar and wind. One of the risks is that it doesn't blow when you need it to. The sun doesn't shine when you need it to. One of the most urgent reliability questions for a utility and a consumer is electrical supply in times of crisis, during hurricanes, during earthquakes. One can't guarantee that solar and wind is going to be there during those kinds of things. The question only makes my point of the complexity of the situation – there isn't one single answer, either for one utility or one city or one country, nor is there one right answer. We're all learning that, as we dig into this issue, solutions are going to be designed to be optimized around a particular problem in a particular city in a particular environment, both regulatory, economic and national.

21

TRANSPORTATION

Wesley Warren

The White House Council on Environmental Quality
360 Old Executive Office Building
Washington, DC 20501

In previous positions I have had, I specialized in certain policy areas on energy and environmental issues and found that over time, I was really learning more and more about less and less with the result that after a while I knew almost nothing. Now that I work in the Office on Environmental Policy and do a broader set of energy and environmental issues, I find that I am really learning less and less about more and more with about the same result.

So I am not here today to claim that I am an expert on transportation policy. However, my office, The Office on Environmental Policy, does have a responsibility for general tracking of progress and implementation of the Climate Change Action Plan, and transportation is a very important part of that.

In the Climate Change Action Plan, there are four action items devoted specifically to the transportation sector. Those are parking cash out, a transportation strategy, promotion of telecommuting, and tire labeling. Altogether these action items would yield about an eight million metric tons of carbon equivalent reduction out of a total action plan reduction goal originally of 109.

On the other hand, the transportation sector is nearly a third of greenhouse gas emissions in the United States. So a number of our critics have said, "What is going on in the transportation sector? How come it is such a large part of the contribution of the total amount of emissions, and yet it has such a small part in the action plan?"

So I would like to first of all to discuss our thinking in that area, the challenges we had to deal with, and why the action plan looks the way it does in this particular area. In respect to transportation policy, there are basically three ways to reduce emissions from the transportation sector. You can reduce the number of miles that people drive, you can increase the efficiency of the automobiles with which they drive, and you can change the kind of fuels that they use to power the automobiles that they drive.

The difficulty is that there are millions of point sources in this area, more than any other area except perhaps the agriculture section. It is a much more diffuse problem, because every time you get in an automobile to drive somewhere, you are a point source.

There are also very complex interactions in this area. It would be nice to say just increase fuel efficiency, and as a consequence of that, we will have a lot fewer emissions. But you find when you lower the price of traveling, whether you do that by actually lowering

Evaluating Climate Change Action Plans, edited by White
Plenum Press, New York, 1996

the cost of gasoline or increasing the number of miles you get per gallon, there is a tendency in the system to increase the number of miles that people are willing to drive.

In addition to that, there are very long lead times in many transportation policies, and there are a combination of reasons for that. First of all, in some cases, it requires a lot more research and development of technologies that do not exist now, especially in the fuel efficiency area. Furthermore, it is not only development of the technologies, but also once you develop the technologies, you have to get it into the model years, which are often engineered years in advance.

And so although improvement in technology is a very promising area for emission reductions eventually, in terms of getting it within the time frame required by the Climate Change Action Plan, which was the year 2000, most of those emission reduction impacts would be in the post-2000 period.

And finally, reducing emissions requires changing behavior, and this is one of the most difficult things — getting people to change the way in which they do things. America is basically a nation of automotive centaurs, and by that I mean we that we are cars from the waist down. People love their cars. They basically view cars as an extension not only of themselves but of their living rooms. Frankly, you have all had this experience yourself, when you sit in the front of your car, this is your space. It is a privacy realm, and it's an extension of yourself.

And as long as gasoline is cheaper than bottled water, it is going to be difficult.

In the transportation area, we really need a long-term comprehensive approach. Even though the Climate Change Action Plan tried to identify the areas in which we could have the most significant impact in the near term, everyone should keep in mind that addressing transportation emissions is really going to require a long-term commitment and changing a lot of ways in which people do things.

When we issued the Climate Change Action Plan, we did, at the same time, initiate another process that would address the transportation sector's emissions in the post-2000 period. We singled out transportation especially. A federal advisory committee started meeting in October, consisting of 30 representatives from a wide range of sectors, industry, labor, environmentalists, state and local governments. They are due to issue a report in 1995, a year from when they began, to make recommendations in those three areas of vehicle miles travelled, fuel economy and alternative fuels.

Also, the Administration has other policies that are not part of the Action Plan, such as the Clean Cities Program, in which the federal government works with cities to try to get them to increase their use of alternative fueled vehicles. Currently, we have 25 cities participating with a projected estimated total of 300,000 vehicles – another area in which we could see additional results.

In those four areas, I think the largest overall gain would have been the so-called Parking Cash Out proposal. This proposal would change the tax subsidy which is currently made available to people as part of a perk of working, where you get free parking, which is estimated at a value below $155. Currently, if you are given free parking at work but its value is above $155, you would be taxed. If you are below $155, you pay nothing.

Our thinking in this particular area was to try to create greater choice for the employees by saying: if you are below that $155 level, why not have the employer offer you an option of either taking the free parking or taking assistance to travel in some alternative mode, especially mass transit and, if you take the mass transit option, you would get the cash value of your parking allowance to use towards that purpose.

We combined that concept with several provisions that would have eased the burden on business by allowing them to deduct the cost of administering the program from their taxes, exempting employer-owned parking, and a small business exemption for businesses with fewer than 25 employees.

In addition to this, taking the cash to take mass transit would become a source of income to individuals, which would, of course, be only taken on a voluntary basis. It would have yielded a substantial amount of tax revenue, about $2 million over five years. We were hoping that this generated revenue for the federal government under a voluntary program would pass last year. We had targeted the GATT, which needed budget offsets, but we did not succeed in getting it in the GATT. Currently, we are reworking that proposal, and we still hope that some form of it will be acceptable to the next Congress.

The second area is a transportation strategy. I will just mention briefly here this has many different components, and is an attempt to weave together a lot of the current authorities that already exist, especially the Clean Air Act and the Intermodal Surface Transportation Act, the principal authorizing act for the Department of Transportation. It includes a number of public outreach programs or programs where we would work with state and local governments to try to reduce greenhouse gas emissions.

It includes a public education campaign that will work through the National Association of Regional Councils, a joint project between EPA and DOT on the conformity rule, which requires that transportation projects be consistent with the Clean Air Act. It relies on the Congestion Mitigation Air Quality Improvement Program under ISTEA, which is a DOE program. It will improve models that exist in the government that project travel impacts, which both DOE and DOT rely on. It will try to evaluate the potential use of the Intelligent Transportation System, which uses advanced tracking and electronic devices to ease congestion and speed traffic flow, which is another ISTEA provision.

It would more greatly promote the Mobile Partner Program by creating service bureaus for state and local governments, and would have the EPA, in its guidance for the Clean Air Act, promote market mechanisms to give SIP credits.

The next area is telecommuting, where people can stay home to work and decrease their vehicle miles traveled. Here, DOT and GSA have a joint program to promote a federal demonstration in Los Angeles and the District of Colombia and a national awareness campaign.

The final area is tire labeling. The rolling resistance of tires on consumer cars very significantly affects the fuel efficiency of that car, and we discover that when consumers buy new tires for their cars after their original tires wear out, they tend to be 20 percent more resistant. If we can get them to continue to buy tires that are as efficient as the original tires they had, you could reduce emissions by four percent.

That action has two parts. One is a proposal to change labeling on passenger tires to list the rolling resistance, and another is to have the industry voluntarily include standards or rolling resistant labels on truck tires.

These four areas are the major approach of the federal government in transportation. We are attempting to rely on market mechanisms, outreach, existing authorities, and to target those areas in which we can have major impacts in the short term.

22

DISCUSSION WITH WESLEY WARREN

Mr. Fox-Penner: Administration officials have suggested that if the voluntary measures in the Climate Change Action Plan do not deliver the necessary reductions, additional measures, including mandatory measures, will be taken. This implicit threat is supposed to provide an added incentive for industry participation and support, but additional actions are unlikely to receive congressional support. And even with congressional support, there will be little time to implement new measures to yield reductions by year 2000. Without a real threat of further action, aren't voluntary programs seriously undermined? Is failure to participate really 'a vote for taxes and regulation'?

Mr. Warren: I'm glad to get these questions on transportation policy. This is obviously important, and one of the other responsibilities that my office has in tracking progress of the Climate Change Action Plan is to ask what will we do as the implementation of the plan proceeds.

When the Action Plan came out just over a year ago, one of the items that we committed to was to review progress under this plan no later than two years after the issuance of the plan and to propose additional measures, if necessary. The Administration has confirmed its commitment to that statement. That is still our view at this time. We will be conducting that review, and we will be examining additional measures if they turn out to be necessary.

I would state the strong preference of this Administration for voluntary cost-effective programs, especially those that pay for themselves through savings to the consumers and create jobs. There's been a lot of talk about what the election results mean in terms of control of both houses of Congress by the other party, but we would at least like to approach this, as we always have, as an issue that is worthy of bipartisan support. We intend to take that message to Congress, initially to try to persuade them that yes, these are programs that make sense from the point of view of people who promote market mechanisms.

Nevertheless, we'll have to wait and see how that goes with the Congress, and I would like to say that if funding is not available, we are still standing by our commitments.

Mr. Fox-Penner: Degassing mines is critical to mine safety programs. The economics to capture and utilize coal bed methane is the limiting factor. Please discuss coordination with DOE and EPA on the coal bed methane program.

Ms. Hogan: We know that degassification is important for safety, and at most coal mines a large portion of the methane emissions come out of the ventilation systems in a concentration of one percent or less in air. We don't view that stream of emissions as something you can economically capture and reduce at this time. However, at the gassier mines that use long wall mining, they have gob wells in the roofs of their mines where they are pulling a medium quality gas, 50-percent methane or more. The mines down in Alabama, Black Warrior Basin, have actually been successful in pulling out gob gas of 90-percent methane or more and have been able to sell that to a pipeline system with minimum treatment.

At the gasiest mines in the country, some 50 to 75 mines, there is a valuable commodity to pursue, and we've done an economic analysis that shows that the economics are there. However, there are a number of other important factors that must be considered. One of them is the ownership issue. Until EPACT was passed, the Appalachian states had very unclear ownership rights established for the gas. In part, the gas rights were thought to belong to gas companies, the coal rights to coal companies, and no one knew who owned the gas trapped in the coal. So nobody was going to make an investment to get the gas if they didn't think it was going to be theirs, particularly because when you trespass on somebody else's resources, there's a triple damages problem. Not only would you lose the gas, you could be subject to triple damages.

The property rights issue has been addressed by the State of Virginia. They acted on it themselves. EPACT put into place a system to address these ownership rights, and gave them three years to address it at the state level. If the states do not pass their own legislation they're going to be asked to adopt a federal model. The Department of Interior is putting this rule together.

In the near term, we can expect that the ownership issue will be addressed, and that the coal companies will be more interested in looking at this methane. Since Virginia passed their law, there has been a lot of activity in drilling wells and in building the necessary pipelines you would need to get that gas from the coal mines to the nearest connection point. The economics really are there. You just have to get over these barriers.

We coordinate very well with the Department of Energy on this program. You may be aware that the Department of Energy has a related program on the R&D for coal bed methane, and we are in constant coordination with them.

Mr. Fox-Penner: In developing countries, what are the relative opportunities for methane reduction by coal gas recapture versus ruminant productions? And what is EPA's involvement?

Ms. Hogan: There are a lot of opportunities in both of those areas. In recovering methane from coal mining, there are a number of opportunities in countries such as Poland, Czechoslovakia, Ukraine, Russia, China and India. And we actually have programs going in each one of those countries, working to outline potential projects and getting those outlines into the hands of the right international lending institutions to get real demo projects going and establishing clearing house networks so that we can overcome informational barriers.

With ruminant livestock, there is another great opportunity for methane reductions. We've done feasibility studies for enhancing dairy productivity in countries such as Nepal, Bangladesh, China, India, Tanzania, and Brazil. And in each one of those countries, there are cost-effective ways to provide feed supplements to the animals to increase their milk productivity while reducing methane sizably.

23

EPA'S PROGRAMS FOR COST-EFFECTIVELY REDUCING METHANE EMISSIONS AND EMISSIONS OF OTHER GREENHOUSE GASES

Kathleen B. Hogan

Methane and Climate Branch
Office of Air and Radiation
Global Change Division (6202J)
U.S. Environmental Protection Agency
401 M Street, SW
Washington, D.C. 20460

Methane is a major greenhouse gas, second only to carbon dioxide in its contribution to future global warming. Methane concentrations have more than doubled over the last two centuries and continue to rise annually. These increases are largely correlated with increasing human populations. Methane emissions from human related activities currently account for about 70 percent of annual emissions. Because methane has a shorter lifetime than other major greenhouse gases, efforts to reduce methane emissions may fairly quickly be translated into lower atmospheric concentrations of methane and lower levels of radiative forcing. This fairly quick response would have the benefit of slowing the rate of climate change and hence allow natural ecosystems more time to adapt.

Importantly, methane may be cost-effectively reduced from a number of major sources in the United States. Methane is a valuable fuel, not just a waste by-product, and often systems may be reconfigured to reap the fuel value of the methane and more than justify the necessary expenditures. Such options exist for reducing methane from coal mining, natural gas systems, landfills, and animal management. These options have been implemented to limited degrees. However, there are a number of barriers that hinder the more widespread use of technologies, including institutional, financial, regulatory, informational, and other barriers. This paper describes the programs that EPA has underway as part of the United States Climate Change Action Plan to overcome these barriers and profitably attain over 15 MMT in carbon emissions reductions by the year 2000. The paper also describes efforts to profitably reduce emissions of other potent greenhouse gases, perflourocarbons.

Evaluating Climate Change Action Plans, edited by White
Plenum Press, New York, 1996

BACKGROUND: METHANE IS AN IMPORTANT GREENHOUSE GAS

Methane is an important greenhouse gas and a major environmental pollutant. Methane is also the primary component of natural gas and a valuable energy source. Options for reducing emissions of methane into the atmosphere offer one of the most effective means of mitigating global warming in the near term for several reasons:

- **Methane is one of the principal greenhouse gases**, second only to carbon dioxide (CO_2) in its contribution to potential global warming. In fact, methane is responsible for roughly 18 percent of the total contribution in 1990 of all greenhouse gases to "radiative forcing," the measure used to determine the extent to which the atmosphere is trapping heat due to emissions of greenhouse gases.[1]
- **Methane's atmospheric concentration is rising rapidly**, more than doubling over the last two centuries and currently increasing at about 0.6 percent per year,[2] in contrast to CO_2, whose atmospheric concentration is increasing at 0.4 percent per year.[3]
- **Methane is a potent contributor to global warming**. On a kilogram for kilogram basis, methane is a more potent greenhouse gas than CO_2 (over 60 times greater after 20 years and about 25 times greater after 100 years).[4]
- **Reductions in methane emissions will produce substantial benefits in the short-run**. Methane has a shorter atmospheric lifetime than other greenhouse gases — methane lasts around 10 years in the atmosphere, whereas CO_2 lasts over 200 years.[5] Due to methane's high potency and short life span, stabilization of methane emissions will have a rapid impact on mitigating potential climate change.
- **There are many opportunities to modify human activities that emit methane because human related activities account for about 70 percent of annual methane emissions**. Methane is emitted into the atmosphere largely by anthropogenic (human-related) sources, which currently account for approximately 70 percent of the estimated 500 teragrams (Tg) of annual global methane emissions.[1] Anthropogenic sources of methane emissions include: natural gas systems; coal mining; landfills; domesticated livestock; liquid and solid waste management; rice cultivation; and biomass burning.
- **In contrast to the numerous sources of other greenhouse gasses, a few large and gassy facilities often account for a large portion of methane emissions**. Applying emissions reductions strategies to these gassiest facilities would result in a substantial decrease in estimated current and future methane emissions levels.
- **Because methane is a source of energy as well as a greenhouse gas, many emissions control options have economic benefits**. Methane emissions are usually an indication of a system inefficiency. In many cases, methane that would otherwise be emitted to the atmosphere can be utilized to generate power or the quantity of methane produced can be significantly reduced through the use of cost-effective management methods. Therefore, emissions reduction strategies have the potential to be low cost, or even profitable. For example, methane recovered from coal mines, landfills, and animal waste systems can be used as an energy source, and options for reducing methane emissions from livestock also improve the productivity of each animal.
- **Well demonstrated technologies are commercially available for profitably reducing methane emissions**. For the major sources of anthropogenic methane emissions (except rice cultivation and biomass burning), cost effective methane reduction technologies are already commercially available.[6,7] Additionally, a

number of other technologies are under development. While offering substantial emissions reductions and economic benefits, these technologies have not been implemented on a wide scale in the U.S. or globally because of financial, informational, and institutional barriers.[8]

- **Reduction in methane emissions provide other important environmental benefits.** Reductions in methane emissions will provide other important benefits for the atmosphere such as reducing the risks of increasing tropospheric ozone and reducing the earth's oxidizing potential.[1,6,9] That is, stabilizing CH_4 concentrations may reduce expected global tropospheric O_3 increases, and while more uncertain, may allow OH to return to about current levels by 2100 after a significant suppression.[10]

EPA's Methane Programs

The U.S. EPA has five programs under the United States Climate Change Action Plan designed to reduce methane emissions from the major methane sources in the United States. In general, EPA's approach to reducing U.S. methane emissions involves a close working relationship with the relevant industries, with the aim of identifying and encouraging implementation of economically attractive ways to capture and use methane as an energy source. Some of these initiatives are voluntary programs for which companies sign a memorandum of understanding with EPA, whereby they agree to identify and implement profitable methane control strategies. Examples include the Natural Gas STAR program, under which the natural gas industry agrees to adopt a core set of emissions control technologies wherever profitable, and the AgSTAR program, under which livestock producers agree to capture and utilize the methane generated from livestock manure, wherever profitable.

Other programs are designed to overcome regulatory, information, technological, and other barriers to implementation of methane control and utilization options. For example, the Coalbed Methane Outreach Program includes a number of efforts to encourage mines to view the methane they now recover for safety reasons as an energy resource, including feasibility studies, information gathering and analysis, and an ad hoc industry working group. In addition, EPA is finalizing regulations that will require landfill owners to control emissions of nonmethane organic compounds; to assist landfill owners and operators in profitably complying with this rule, wherever possible, EPA has designed and launched the Landfill Methane Outreach Program to educate landfill owners and operators about the economic and environmental benefits of using their landfill gas to produce energy. Each of these programs is described in more detail below.

Natural Gas STAR. The Natural Gas STAR program is a voluntary public/private partnership that reduces methane emissions by promoting cost-effective technologies and practices throughout the U.S. natural gas industry. This program is based on a good understanding of the technologies in the field that emit methane and an analysis of the technologies and practices that can be changed so that emissions can be reduced. The program currently promotes enhanced inspection and maintenance programs at compressor stations and gate stations, replacing high-bleed pneumatics with low bleed devices, and replacing engines with turbines where technically feasible. Natural Gas STAR provides technical assistance, implementation guidelines, and an information sharing network for gas companies to achieve cost effective emissions reductions. The existing program was launched in March of 1993, and now includes over 35 transmission and distribution companies representing over 50% of transmission pipeline mileage and 35% of connection services. The program will be expanded to production companies in Spring 1995. The program is expected to provide 4.3 MMT of carbon reductions by 2000.

We Have a Landfill Methane Outreach Program. There are over 6,000 active landfills in the United States. Landfill Rules being promulgated by EPA to control air pollutants from landfills will affect a fraction of these landfills, about 10 percent, and provide substantial reductions in methane emissions. Further reductions in greenhouse gas emissions can be achieved by promoting energy recovery by affected landfills rather than the minimum requirement of flaring the landfill gas. The Landfill Methane Outreach Program provides necessary information for overcoming the barriers to profitable landfill gas recovery. The Outreach Program targets landfills affected by the Rules as well as landfills not affected by the Rule that could profitably recover and utilize the methane they emit. This second landfill group has not installed energy recovery systems because they also face a number of barriers, such as disincentives for utility purchases of landfill gas, artificially low prices, lack of information, regulatory constraints, and technological constraints. The program includes development of outreach materials, such project development handbooks (including case studies and sample RFPs) and landfill profiles to identify the most attractive project opportunities. The program works directly with utilities and states to overcome key informational, regulatory, and other barriers. The Landfill Methane Outreach Program was launched in December 1994 with 8 State Allies and 12 Utility Allies. It is expected to provide 1.1 MMT in carbon reductions by 2000.

Coalbed Methane Outreach Program. The Coalbed Methane Outreach Program promotes opportunities for profitable emissions reductions with key coal companies. The outreach program targets 50 to 75 of the country's gassiest mines that currently recover significant quantities of methane for safety reasons, but vent it to the air because of property rights issues or because they have not identified a use for it. The Program includes development of outreach materials such as mine profiles that identify attractive use for the gas being released from the gassy mine and performing briefings for companies, states, utilities, and others. It is important for many to know that the reductions that can be achieved at one mine would be all the reductions that many utilities need to fulfill their Climate Challenge pledges. The Outreach Program will also help ensure that the Energy Policy Act provisions resolving coalbed methane ownership issues in key states are effectively implemented so that the full range of potential reductions can be achieved. This program can lead to thousands of jobs in the Appalachian regions while also providing 2.2 MMT of carbon reductions by 2000.

AgSTAR. AgSTAR is a voluntary pollution prevention program with the livestock industry. The AgSTAR program encourages dairy and swine facilities to adopt animal manure best management practices that profitably manage animal manure, protect surface and groundwater, and reduce fertilizer costs. The AgSTAR program overcomes a number of barriers which currently hinder the more widespread use of on-farm energy recovery technologies. These barriers include farmers' lack of information about the profit potential of methane recovery options and their memory of projects launched during the 1970s that failed. Current technologies overcome problems of the past, including now being less expensive to design and maintain. Under the program, producers commit to survey their facilities to identify profitable options for capture and use of methane for on-farm power usage and then to install these technologies where profitable. This program is expected to provide 2.3 MMT of carbon reductions by 2000.

Ruminant Methane Productivity Program. Dairy and beef cattle are responsible for over 30 MMT of methane emissions per year. This program will address the six main options for reducing these emissions through improved management at the farm level. These options are as follows: (1) improved nutrition through mechanical and chemical feed processing, (2) improved nutrition through strategic supplementation, (3) production enhancing agents to improve feed efficiency, (4) improved production through improved genetic characteristics, (5) improved production efficiency through improved reproduction, and (6) controlling disease. The program will also build on existing efforts to remove market barriers to and create incentives for increased production of milk and meat with lower fat content. Such products cause less methane per unit of product produced than higher-fat products. This program is expected to provide 1.8 MMT in carbon reductions by 2000.

EPA's Program for Reducing Emissions of Perflourocarbons

EPA has also developed a program to reduce emissions of perflourocarbons, CF_4 and C_2F_6, from primary aluminum smelting in the United States. These gases are highly potent greenhouse gases, thousands of times more potent at trapping heat in the earth's atmosphere than carbon dioxide. EPA's program, entitled the Voluntary Aluminum Industrial Partnership, encourages aluminum producing companies to reduce emission of CF4 and C2F6 where technically feasible and cost-effective. Because the factors that cause these emissions result in an efficiency loss, a reduction in these emissions will improve smelter efficiency. This program includes support of research efforts, improving the understanding of emissions from different smelter technologies, and developing flexible plans with the industry. Emissions of CF4 and C2F6 may be reduced by 30 to 60% through voluntary efforts of the aluminum industry. This program is expected to provide 1.8 MMT of carbon reductions by 2000.

SUMMARY

In summary, methane and other greenhouse gases present substantial opportunities for profitably reducing greenhouse gas emissions in the United States and play a large role in the United States Climate Change Action Plan. EPA's programs to profitably reduce the emissions of these gases are strategically designed to overcome the barriers which have inhibited otherwise profitable projects, and these programs are on track to deliver the carbon reductions put forth in the Action Plan. Furthermore, there are many opportunities in other countries for reducing emissions of these gases profitably—opportunities that are attractive to US providers of goods and services.

REFERENCES

1. IPCC, *Climate Change: The IPCC Scientific Assessment*, Report Prepared for Intergovernmental Panel on Climate Change by Working Group 1, 1990.
2. L.P. Steele, E.J. Dlugokencky, P.M. Lang, et al., "Slowing Down of the Global Accumulation of Atmospheric Methane During the 1980's," *Nature*, No. 358: 313-316, (July 23, 1992).
3. Oak Ridge National Library/The Carbon Dioxide Information Analysis Center, *Trends '91*, Atmospheric and Climate Research Division, U.S. Department of Energy, Oak Ridge, TN, 1992.
4. IPCC, *Radiative Forcing of Climate Change: The 1994 Report of the Scientific Assessment Working Group of the IPCC*, 1994.
5. G.L. Vaghjiani and A.R. Ravishankara 1991, "New Measurement of the Rate Coefficient for the Reaction of OH with Methane," *Nature*, No. 350: 406-409 (April 4, 1991).
6. K.B. Hogan, J.S. Hoffman, and A.M. Thompson, "Methane on the Greenhouse Agenda," *Nature*, No. 354: 181-182 (1992).
7. U.S. EPA, *Options for Reducing Methane Emissions Internationally, Volume I—Technological Options for Reducing Methane Emissions*, Report to Congress, K. B. Hogan (ed.), Office of Air and Radiation, Washington, D.C., December, July 1993.

8. U.S. EPA, *Options for Reducing Methane Emissions in the United States*, Report to Congress, K. B. Hogan (ed.), Office of Air and Radiation, Washington, D.C., 1993.
9. WMO, *Scientific Assessment of Stratospheric Ozone: 1989*, Report No. 20, World Meteorological Organization Global Ozone Research and Monitoring Project, Geneva, Switzerland, 1990.
10. A.M. Thompson, R.W. Stewart, M.A. Owens, et al., *Atmospheric Environment*, 23: 519-532 (1989).

24

FUNDING THE U.S. CLIMATE ACTION PLAN

T. J. Glauthier

Natural Resources, Energy and Science
Office of Management and Budget
Old Executive Office Building, Room 246
Washington, DC 20503

It is nice to be back with this group. I have enjoyed being associated with this conference now for the last few years, and was impressed to see on the program that this is the tenth anniversary of the conference. It really does add some perspective to the issues and to CEI.

I was also pleased to see on the program people like Jim White, who I worked with a couple years ago in the steering committee on the conference, and of course Liz Thorndike, who has led this program for a number of years as its founder. And I think this has been a really good conference, every December, bringing people together in Washington on this issue. When I was tracking the issue before I got into the government, it seemed to be a key time to be able to hear from the people who were involved from the different agencies and the other interest groups affected by the issue and get a chance to get an update. So I'm really pleased to be able to be here and be part of it.

I was also interested to see how many familiar names there are on your steering committee. Unfortunately, a lot of them are not here at the moment, but I was running down the list and saw Chris Bernabo, Alex Cristafaro, Roger Dower, Howard Gruenspecht, Rich Richels, and Carol Werner; a whole lot of the people that I have dealt with on this issue over the last several years.

As you said, too, I was working on this with the World Wildlife Fund, and it was just two and a half years ago that we were at the Rio Earth Summit and got the treaty signed. It seems sometimes like it was a long time ago,just two and a half years, and a lot has happened in that period.

And it was just four years ago in October that we had the Second World Climate Conference in Geneva, which was the beginning of the negotiations for the treaty. So in some respects, an awful lot has happened in such a short period of time.

In regard to trying to add some perspective to these various issues, the election on November 8 has led a lot of us to do some thinking about perspective and how to put all of the changes in context. The statement that I found seems to explain what happened is this one. It says, "The trouble with the Democrats up to now has been that they have been giving the people what they thought the people ought to have, rather than what the people wanted." Well, that's not very profound, but it's interesting that it was stated by Will Rogers in the 1920s. So things haven't changed all that much.

Evaluating Climate Change Action Plans, edited by White
Plenum Press, New York, 1996

I find turning back to Will Rogers is helpful as one of the more insightful commentators on the political scene. He was the one, of course, who said, "I'm not a member of any organized political party. I'm a Democrat." He also said, "The greatest trait to recommend to Democrats is optimism and humor. You have to be an optimist to be a Democrat, and you have to be a humorist to stay one."We're working at it and intending to stay, but it takes a little humor.

I'd like to make some comments about where we are in the Climate Change Program, how we've arrived here and what we're doing and then to open it up for some questions, so that I can be sure to touch on things that may be of particular interest to you.

I want to talk first about the plan that we had a year ago. You've all seen the plan announced last fall, and that plan has been described here, I'm sure. I note that you will have papers from several people who really have been actively involved in the plan, both its development and now its implementation.

I don't want to repeat a lot of things, but I think it's really important to think about the nature of the plan that we put together and the fact that there's a real commitment to having a plan, having actions that are laid out to meet this goal of returning our emissions to the 1990 levels by the year 2000, and doing it in a way that is clear and specific. It's important that the plan is based as much as possible in ways that are win-win actions, actions that do benefit our industries, benefit jobs, work with the economy, things that can be done in flexible ways, so that we truly marry jobs and the environment. These should not be in conflict, in fact, jobs and environment issues can really work together very well. Our commitment is that, in the long term, the way to have a healthy economy is to be sure that we're making these changes in the right direction here.

The process of developing the plan last year was not one that we want to repeat. There were a lot of meetings. I think one of the characteristics of the early part of the Clinton Administration has been endless rounds of meetings. It was good in some regards because we got a lot of input but it took an incredible amount of time.

At OMB is we tried to take the role of scorekeeper. We put together a series of tables that could be used to organize what all these suggestions were and what they would accomplish, what all the elements of the action plan would be. We started with something on the order of 150 different actions and ended up with about 50 in the final action plan.Our effort was to be very specific about what those actions were going to be, who was going to be responsible for them, what we were expecting to get out of them, and what resources had to be put in,in terms of dollars and people and what time phasing was involved.

We tried to be specific so that we could really be sure we were delivering the resources in order to make those commitments. Now we're involved in the process of trying to track those,to be sure that the agencies are following through and living up to the program that we all committed to.

There's a range of activities, of course, as you've seen, looking at the plans and hearing the other speakers. And many of you are probably involved in different elements of this plan. They are very broad ranging, from the programs that preexisted, like the Greenlights Program or some of the research programs, to additional efforts in the areas of buildings and motors, the utility Climate Challenge Program, the effort to bring in utilities and get them involved in a very direct way, and finally to elements that actually could generate revenues.

A key feature in that last category is the proposal we made last year for the parking cash out actions, which we tried very hard to get into the law so we could develop that as a feature that would both help us in terms of reducing driving, reducing emissions, increasing the use of mass transit, and had a side benefit of actually generating revenues for the government. We haven't gotten that into law yet,we're still working on it.

As we went through the funding last year, developing the 1995 budget, we proposed a very aggressive set of increases, for renewables, for energy conservation and efficiency, for other specific programs and different agencies. Overall it added up to $328 million that we were asking for as an increased level of funding over and above what the base programs were before this.

We worked very actively with the Congress trying to get that whole program implemented. We spent a lot of time meeting with the key appropriations subcommittees, especially the Department of Energy and EPA. We worked with people on the Hill to explain the programs to go through them in detail so that it just didn't seem like big numbers.In fact the conservation program is made up of a lot of different actions, different programs. It's crucial to explain what was going to be done, why that was justified, how it related to the Energy Policy Act or how it related to other programs as well so that it wasn't all focused on climate change.

That will be one of the themes that we're certainly going to push with the new Congress, that there are more and more reasons to do these things than just climate change as a single justification.

The Vice-President sent a letter to the appropriators last spring urging full funding for these climate change programs in the budget. The result was we got about two-thirds of the money that we had asked for. We got differing amounts, of course, in different program areas, but we did well overall. We did very well on renewals and energy programs We had asked for such a big increase in the conservation and efficiency funding at DOE that we did not get all of that money. It had to do with a number of other problems in that appropriations subcommittee which has the Indian Health Service, the whole Department of Interior, as well as those programs. And we had asked for a 42-percent increase there. We got a 15-percent increase, which was a very healthy increase. And yet, it wasn't up to the total dollar magnitude that we wanted.

We've got to request the additional money now. One of the reasons that they didn't give us the whole increase is they weren't sure we could actually ramp the program up that fast and be effective.

In terms of implementation, we've been working over this last year to actually carry out these programs at the same time that we were trying to get the additional funding for the next year. We have 40 of 47 action plans, or action plan activities, up and running right now, and we feel there's good participation, good support from the business community,and good support from the governmental units.

Examples might be things like the electric utility Program, the Climate Challenge Program, which now, I understand, has over 800 electric utilities signed up. It's got 80 percent of the U.S. electric generating capacity in the program. This is a way to try to get the utilities working in programs in which they can develop a good deal of flexibility to meet overall goals.

We've got commitments that should lead to reductions of ten million tons of carbon dioxide by the year 2000 from that program. It's one where we're trying to be flexible,to work with utilities on a voluntary basis, but also be specific, get some commitments, do some tracking, see what happens. Because if it doesn't happen, frankly, we've got to come back with something else and that, I think, is the reason that many of these private sector organizations are participating. Some of you may represent those organizations..

We re expecting similar results in the Green Lights Program where participation is up 30 percent over last year, and the CO_2 reductions expected there are 13 million tons over the next five years.

You may have heard of the various green programs. There's also the Golden Carrot Program, which is for new technologies. One program that's gotten some publicity is the Refrigerator Program, to try to get a super efficient refrigerator. The first one of those has been developed and installed at the White House now. That contract was won by Whirlpool and is an example of efforts to try to stimulate developments in areas that will have some commercial benefits and do more than just reduce emissions. It will also advance us in technology areas.

I understand now that there are a lot of participants in the Natural Gas Star Program, which has about 30 gas corporations representing 50 percent of the transmission company pipeline miles in the country's gas network. We're trying to involve those people in voluntary programs that will help us on natural gas and natural gas emissions.

OMB's role has been to try to help facilitate and support these programs. We're not trying to be in charge. We're not an operational agency, but we sure can help in coordination and in follow-through, where agencies are trying to do something, but they run into a barrier one way or another. We can sometimes be a big help in approaching some authorities. We can make sure that the funding is there, in particular, where there are programs that across several agencies. We can ensure that they actually are getting support consistently across those agencies.

We're dealing with areas of interagency cooperation, where agencies ought to be working together but have their own agendas sometimes, and there's a little turf consciousness. We try to see what we can do to try to bring them together to a common goal and a common objective.

We also are trying to do the review that I described earlier, to compare what the agencies are doing against that original plan. Are we actually carrying out the actions that we talked about? Ultimately, we want to get to reviewing the results. Are we getting the emissions reductions that we expected? It's a little early for that. At this stage, we need to know whether we're taking the steps that we had planned. Are we doing these things? Are we doing them at the levels we anticipated? What are the problems that may be slowing it down, if it's not going forward?

We also have the responsibility to coordinate legislation within the Administration. Any new legislation or legislative initiative or even any agency's position on legislation has to come through OMB for clearance, so that the agency is speaking on behalf of the Administration and not just from its own perspective. That gives us the ability to be sure we get the different agencies together on legislation.In this area, legislation tends to involve more than a single agency often, too. So we're working actively there.

I don't know if you've heard about the activities of Post-2000 Planning. That's another area where we have to look beyond just meeting the target in the year 2000, obviously. We've got to be doing things now that are going to lead to long-term emission reductions.

The most important areas there are probably the ones that link research and development on longer term technologies to uses of energy. Transportation programs are particularly important there because we've got to do things involved in the replacement of our present fleet of vehicles. Changing patterns of behavior will take a long time to really show up in a measurable way, but this is essential to make the long-term kinds of changes that we need.

The Clean Car Initiative is a most visible example and is something that we are really strongly committed to, something that all three auto companies have signed on to. It's a good example of, again, the effort to work with people in the industries and to try to harness our resources in the national laboratories and in the different departments, the Department of Commerce,and the Department of Energy. Even the Defense Department has got parts of that program.

That program, I assume you know, is to try to triple the average mileage of a vehicle in the next ten years, so that we will get to the kind of step function increase that really will make a difference. It involves redesign of virtually all elements of the vehicle and the vehicle system.

We're also trying to use the government's purchasing power where we can, and we're seriously committed, for example, to using the government's fleet purchases to get alterative fuel vehicles into federal fleets and on the road now. The cars we can get now will be based on natural gas or be methanol/ethanol fuel-based vehicles. There are things that we can do here and now and, as the technologies improve over time,we will continue to move forward. We expect to have purchased over 20,000 alternative fuel vehicles by the end of 1995 and another 10,000 to 15,000 next year in 1996. We're really making a serious commitment, trying to move ahead, and hope to stimulate the development and commercialization of these technologies.

We're also into buildings. The building that I'm in, the Old Executive Office Building, has had a number of improvements, the windows were replaced about three months ago with thermopane glass. Previously it was a single pane glass that often didn't fit well and let an awful lot of heat out in the wintertime and cold air out in the summertime. Unfortunately, they haven't got it exactly right because the window glazing is also tinted, and the windows now are darker. I used to be able to get by with one bank of lights on in my office. I have three banks of lights now that are burning. It's a little bit counter to the overall goal of reduced energy usage but I think we're probably ahead, energy wise. I understand that we could have had clear glazing if we had specified that properly in the contract, but didn't think it through.

In both the White House and the Old Executive Office Building we started an energy audit about a year ago and in all the hallways now we have compact fluorescent light bulbs everywhere. We're doing a lot to try to improve things.

You will still find window air conditioners in my office because it's a solid granite building,and we can't quite figure out how to put in central air conditioning. They did come around six weeks ago and put controls on the radiators, so I can actually turn the radiator off. The old ones were painted over so many times in the 100 years that that building has been there that you couldn't adjust anything. In the wintertime there were some offices, mine occasionally, that even had the window open. We're getting better, but it will take awhile.

We are encouraging all of our agencies to participate through the Federal Energy Management Program, the FEMP program, and to try to do things that will really move them forward. Our office can be an example if we do the right things, but obviously, it's just one office building.

Throughout the whole federal government system, we've got a lot of office buildings, and we are trying to structure things so that we can use the savings from contracts and the energy costs to provide incentives, both to pay for the contracts and to provide some incentive back to the offices that sponsor the contracts. We're trying to do things that will have a lot of leverage and, where we can, set up new authorities to do these things more easily, more flexibly.

You may have seen some articles about the National Gallery and the heating and ventilating system there; the problems that they're having with the artwork there, and questions of whether or not there is artwork damage. I'm glad to say that we were actually ahead of the curve a bit on that one, and we have arranged to install a new high-efficiency system, that the Department of Energy is sponsoring, and will go in there sometime in the next year. With the number of visitors that the National Gallery gets, we hope that we can get some publicity to use as leverage on the overall point of more efficient, better systems.

Looking ahead to next year is the topic that I was really supposed to talk about. What's going to happen with the new Congress? What's going to happen with the budget that we are working on now and will be sending up to the new Congress?

Up until November 8, we were working along very well on the budget, and we had continuation of the existing programs pretty well all laid out. In this post election period we are going back and reassessing. In fact what we're going to do is going to be very consistent with the themes, with the priorities, and the programs that this Administration is already committed to. The action plan is there because this is a serious problem, a serious risk to the environment, to the U.S., to the world. That risk hasn't changed after November 8. That's still there.

The programs are still needed. The action plan elements are still needed. You can expect to see from our budget a continuation of the same kinds of programs, the same kinds of priorities and those that were laid out last year were the first step in the process. This is the second year. Some of the programs need to continue to increase in order to be on the track that they were anticipated to be on. And that's what we'll be doing.

Will we make some changes because of the November elections? Sure. We will around the edges. We're going to try to make sure that things really stand the test of practicality, that there are multiple benefits from these programs, and demonstrate that these programs really work for industry's and the environment/s benefit, and possibly have export market potential as well.

But those are already characteristics of the plan, so it's not like we're changing things a lot. It's more a question of making sure that we've really got things in order, we're doing the right things. Then we're going to push them along.

Research and development is another area that is part of the plan that has to be continued. It doesn't have as immediate short-term payback as do the other actions, in terms of emissions reduction or even in terms of commercial technologies, but it's essential. We have had a commitment from the Administration to science and to R&D, and that's a commitment that will continue. I think you'll see that theme in many areas. Things like the Clean Car initiative will illustrate that.

We would like to have your help as we go forward in this next Congress. It's going to be difficult, not only because we have changed the leadership of both houses, but we also have so many new members. A third of the members of the House of Representatives are new members in Congress. There's just a lot of education and understanding of these programs needed from all sides.

So I'd ask your help in working with the members of Congress and with others in town to understand why we're doing the things that we are, what the benefits are, what the needs are for these programs.

We know is that if we label these programs in broad ways and just say, "This is what we need for climate change," that that doesn't work, they need to understand the specifics. Why are we doing this Motors Program? Why do we care about the efficiency of electric motors? It turns out a tremendous amount of electricity is wasted every year because of the inefficiency of our electric motors. That isn't very visible but if you get down to specifics, to the real benefits to the economy in making those kinds of changes, that's something that people can support. I think it's those kinds of things that we were making a lot of headway with last year. We need to do that as much and more this year.

We are going to need to continue, on our side to lead by example, to do what we can in the government's energy use; the government's buildings, its fleets and the like, and in recycling and other behavior. We would like your support as well, and hope that you will follow the example or, in many cases, your leading, and we'll try to follow your lead.

We'd like to identify ways to make these programs work better. We take pride at OMB in trying to help solve problems and if there are programs that should be moving ahead but aren't, or there's cooperation that should be taking place, but isn't- these are things that I particularly feel we ought to be able to solve. We should be the management consultants, if you will, in the government who can help in implementation where there's a barrier of some sort.

Often, it's relatively easy to fix things once they are identified. Once we've figure it out, we can do something about it just because we happen to be positioned in the government at the right place. We've got the leverage there to make it work.

If you see those things, if you're working on programs that are having problems, and it ought to be possible to do something better, tell us. Let us know. Let me know. We'll see what we can do.

I'd like to end on one quote here from Harl'and Cleveland, who was a senior government official a number of years ago. Most of his work was in the foreign policy area, but he made a statement that I think works or applies very well to this situation. He said, "We know in our hearts that we're in the world for keeps. Yet we're still tackling 20-year problems with 5-year plans, staffed by 2-year personnel, working with 1-year appropriations. It's simply not good enough."

Well, we're dealing within a long-term situation here. The climate change problem is not just a 20-year problem; it's a 100-year problem and beyond. We are still working, though, with one-year appropriations.

I'm reminded of a headline that I just saw earlier this week in The Wall Street Journal, dealing with GATT. The headline had to do with the tedious nature of all the little elements in GATT and said that all programs that are significant have their tedious side to them. And I think this is one of those where we have to follow all the little examples, all the little details, through. But in the end, if we do that, if we follow the one-year step, each step of the way, each year, we will get to our goal and will accomplish what we need to for this long-term program.

25

LUNCHEON DISCUSSION WITH T. J. GLAUTHIER

Question: In preparation for your discussions with the new Congress, are absolute numbers changing from what you had done before November 8, or are you simply trying to reestablish the reasons behind these programs in order to sell them to the new Congress?

Mr. Glauthier: We didn't have all the numbers finished, but I would have to admit that there are some numbers that will be different. There are some absolute number changes. I don't think there's a difference in the overall theme or the priority that we give to these programs, but yes, there are some differences. And we're still talking. There may be more that will come.

Ms. Jordan: You mentioned the FY '96 budget for the U.S. Global Change Research Program. Is it separate from the Climate Change Action Plan?

Mr. Glauthier: The two are very related, certainly, both of them are important priorities and we are tracking both of them.

One of the things we have been doing is to run a cross-cut on these environmental issues, these and others, across all agencies, so that we can be sure that EPA, NOAA, DOE, Interior, Agriculture, NSF the others are all in fact doing their parts for the overall puzzle.

Question: How do you expect your proposals on the Climate Change Action Plan to be received by the new Congress?

Mr. Glauthier: Well, they should see the inherent wisdom of these programs. I think that the key, really, is the positioning; that they understand that these are not just climate action change budgets or programs, that these are programs that have benefits for their energy conservation purposes, for our technology development, potential exports and the like; and to the extent we can really help them understand it that way, it will be received well. But that's key, and that's where we need your help in that process.

Mr. Womeldorff: I'd like to go back to your Global Change Research Program for a moment. It's my understanding that you're doing a bottoms-up review which, if that leads to prioritization, would cause interagency transfers of budget authority. What kind of cooperation are you getting from the losing agencies in that? If the total is reduced, then prioritization certainly becomes even more important.

Mr. Glauthier: It is really important. What we're trying to do is deal with what the responsibilities or activities of each of the agencies should be and then make sure we're funding those activities. The end result may be a transfer, if you will, but we're not characterizing it that way. We're not characterizing it as $5 million goes from this agency to that agency. Each of the agencies has so many activities that if we can do this in the context of the annual budget, it becomes more a priority setting process and more identifying what their program activities will be and less a winner-loser exercise.

But it's really crucial that we do it, establish priorities. There just isn't enough money to go around, and it's not going to be easy to justify these programs. They have to really make sense.

Question: You suggested that the plan will be monitored and that if traditional measures are needed, those would be considered. If a Democratic-controlled Congress did not enforce the energy tax before funding the existing program, what hope is there that a Republican Congress will go beyond that and endorse additional measures?

Mr. Glauthier: I think you're certainly right in the macro sense to raise the question. We're looking at this in a perspective that it is going to take us several years to really complete all this monitoring and to make the adjustments. At this point, we're not planning major dramatic changes. We are going to look at what additional steps we need and try to put either more activities into the equation or more resources somehow into the programs that we have.

I'm sure you've talked to people from other agencies about this. At this stage, we're still trying to make sure that all of the current activities are really getting fully implemented, that we're following them through completely. Then we'll see what else is needed.

Question: I wonder if you can give us a specific instance in which you attempt to educate the new members of Congress. Do you have a time line on that or a specific event?

Mr. Glauthier: The way we've done it most effectively has actually been in small settings, in individual briefings and the like, where people would go up from individual offices, like the Department of Energy's office. Christine Ervin's staff would meet with the members of the authorizing committees or the Appropriations Subcommittee, and go through, in some detail, what the elements are of their program and what they are really doing, what they get for it, and why it costs what it does.

That kind of an exchange and more detailed meeting seems to really yield more benefit than more formal briefings. I think we'll do more of that.

That does involve leaving materials behind sometimes that require some preparation, but that's the approach I expect to take and that will be our primary means of going forward.

Question: You started out by saying that you had too many meetings to develop the national action program. Can you tell me what you mean by that?

Mr. Glauthier: Well, no. I was talking internal, not meetings with people in the various communities, but rather just within agencies and the White House complex. The process of getting people together and going through seemingly endless meetings has been raised to a high art form, I think, in the early stages of this Administration.

We've worked better lately, where I think we've had more specialized working groups that will work on specific elements of a program and then come together. I didn't mean to indicate that there was too much consultation with the various public interest groups or the industry. I think we probably will never be able to do as much of that as we'd like to. We really would like to spend a lot of time with people outside the government. We just run out of time, frankly. So on balance, we ought to be doing a bit more of that and a little less of the former.

Ms. Morgan: I was wondering, as the U.S. is formulating its position for INC 11 how much contact do you have with the State Department? Obviously, their ability to agree to something is going to be affected by the domestic situation and the credibility of their plan and I was just curious what the relationship was.

Mr. Glauthier: It has to be coordinated. And in fact for them to make a commitment on whatever position they'll take in Berlin is something that gets reviewed by all the agencies and actually cleared through OMB. We are looking at a range of things already that bear on that, as you would expect. Bob Watson, is very much involved in that at one level. All the agencies and all the major White House offices will be part of that.

Question: This is in regard to international environmental technology cooperation – do you see any new administration policies towards those issues and where it comes in light of gas? And how do you think the new Congress might respond to investments in technology that is climate sensitive?

Mr. Glauthier: Well, I certainly see a lot of continued support for that area. I'm not sure that it's going to be new policy, because I think we already have that degree of support. I think we're going to continue a real strong degree of support for the programs that are leading toward active cooperation with the U.S. industry, ones that we think we can see develop ultimately into export opportunities, business opportunities for us.

I think there will be less of the kinds of programs that you've worked on in the past, where AID is providing money to programs internationally without that focus. That's the kind of thing that the Congress will be looking for, too – they're likely to be less sympathetic to foreign aid and looking more for the purpose of the program. How does it tie in to our industries, to our other objectives and all that? As we can bring that together we will have the strongest base for support.

26

CANADA'S NATIONAL ACTION PLAN

Richard Findlay

Climate Change Group, Air Issues Branch
Environment Canada
351 St. Joseph Blvd.
Place Vincent Massey, 11th Floor
Hull, Quebec K1A 0H3

I know that there are two rules for speakers: that you should keep it short and that you should try to be light and humorous to keep the room awake. I know that the issue of global climate change is not exactly a source of chuckles these days, so the humor part may be a little bit difficult, but I will, indeed, try to make up for that by being brief and allowing some opportunity for discussion afterwards.

One thing that I suppose will aid me in doing so to a certain extent, is by saying that Canada does not yet have a national action plan. I think the key word, though, is yet. We are very close to the point that we will be having one, and therefore I will direct my remarks mainly to those process-related considerations that we are going through, the kind of maze of meetings that was referred to, I think accurately, by a previous speaker.

I would like to begin with a little bit of background. This story begins in 1990 when Canada began the process of development of what we called our National Action Strategy for climate change. This was very much a national process that involved the provinces of Canada and the Federal government.

It developed a kind of policy framework within which we are still working. It has three main components: limiting greenhouse gases, anticipating and preparing for climate change, and improving the scientific understanding of climate issues.

Our work has been based on an approach that involves all sources and sinks of greenhouse gases and the so-called comprehensive approach that has been in place and discussed internationally. And we are working in step internationally through the various processes that are in place, and are in accordance with the 1992 Framework Convention on Climate Change from which all this work flows, and with which I am sure you are quite familiar.

In 1993, the real process began in Canada to develop our national program. We developed a thing called the "National Air Issues Coordinating Mechanism" which is very much a collaborative, multi-stake holder, federal/provincial kind of enterprise. It is a mechanism that reports to all of the Energy and Environment Ministers of all the Federal and Provincial Governments across Canada.

This involves people from business, from environmental groups, from cities, from consumer groups and so on. Another important aspect of this national coordinating mecha-

Evaluating Climate Change Action Plans, edited by White
Plenum Press, New York, 1996

nism is that we are very consciously trying to move in a cross-cutting way. For example, we notice that if you do the right thing for transportation, you inevitably are going to be doing the right thing for climate change, for acid rain, for smog, for hazardous air pollutants and all the other kinds of atmospheric issues together.

We are trying to move from a vertical orientation of issues to a more horizontal orientation for managing purposes.

In November 1993 at a meeting of all the Federal and Provincial Energy and Environment Ministers, we called for this national mechanism to prepare options that will meet Canada's current commitment to stabilize greenhouse gas emissions by the year 2000. And that is Canada's position, that we are going to be in a position to do that.

The ministers also asked us to develop sustainable options, so called, to achieve further progress in the reduction of emissions by the year 2005 with no real targeted number agreed upon nationally. I would note that the Federal Government when elected, stated a 20 per cent reduction target in their policy framework, a book called "Creating Opportunities" that the federal government subscribes to.

In 1994 Canada produced our National Report on Climate Change, basically Canada's report on actions to meet commitments under the Framework Convention. It includes where Canada stands from an emissions point of view, the actions underway, the measures being taken, and the process by which we are proceeding into the future.

This is a snapshot of actions being taken. It meets the international guidelines for reporting, and therefore Canada is in compliance with the Framework Convention requirements.

November 8th 1994, which I know is a significant date in U.S. politics, was also a very significant date in Canadian climate change politics as well. That was the day that the Energy and Environment Ministers from all the provinces and the Federal government got together and reviewed progress over the last year, offered this mechanism to proceed to develop a program, reaffirmed the objective to develop options to stabilize greenhouse gas emissions before 2000,and called for a National Action Program to be finalized and presented at the Conference of the Parties to the Framework Convention in the spring of 1995.

We are working frantically to develop the program.

I would like to say a little bit about Canada's emissions outlook and the challenge that we face in terms of closing the gap. With our energy-related emissions, the gap between 1990 and 2000 is expected to be about 62.4 megatons of CO_2 equivalent, which translates to about 12 percent.

When you include non-energy-related emissions the gap widens to about 13 percent over the same period. And it gets a little fuzzier as we go into the future, as you can expect, but we project by 2005 that this same gap will probably be about 19 percent.

We know we have a huge challenge and we know that Canada is a relatively minor contributor on the world stage – we contribute about 2 percent of world emissions. But on the other hand, we are a fairly intensive energy economy, we enjoy low energy prices relative to the rest of the world and have quite a healthy supply of energy relative to other countries. So, it is a challenge.

I would like to go back to the process itself just for a minute. I think it is important how Canada is proceeding to develop our program. It is very much a collaborative enterprise. It very much involves the stakeholders from business, environmental groups and other sectors. It is kind of an interesting approach, whereby we have a top down imposed structure and mechanism whereby Ministers of the Governments involved, are charged with the responsibility for developing the specific measures in this collaborative process, a kind of a bottom-up thing.

We are trying to do this together with other air issues, and trying to make sure that we can seek efficiencies and synergies by handling all these air issues together.

We have also done to the best of our abilities, I think, a fairly effective outreach process over the last several months where we have had some regional meetings and a fan-out process whereby those participants who are directly sitting at the tables worked with their jurisdictions, whether it be utility organizations or environmental groups or whatever. We are trying to involve as many people as possible in this process.

Just a clarification, at the top of this process are the Ministers responsible for making decisions, but NAISC is actually an acronym for the National Air Issues Steering Committee, and is made up of the Deputy Ministers of all the departments involved, the top level bureaucrats in the various departments.

The NAICC mentioned here is getting a little more operational. We have Assistant Deputy Minister level people involved, as well as stakeholders. There is a stakeholder committee that relates to that level.

The multi-stake holder part of the plan development process begins with the Climate Change Task Group, a group of a couple of dozen people which has been working very hard over the last year. It is completely multi-stake holder and it has been responsible for conducting the outreach process and generating this material.

There are two Working Groups. One is called the Measures Working Group which has worked hard to identify, create and develop clarity around the description of what measures could be; not to decide what they should be, but to make it clear what they could be, and to do a bit of analysis and provide information so that all the people who have a stake in this can better understand where the other participants are coming from, in talking about measures.

The Measures Working Group has done its work. It had created a consensus, a 400-page catalogue of measures, which is available should you wish to have a 400-page catalogue of measures, which has been used as the foundation or the basis for the development of a National Action Program.

The Forecast Working Group has now taken over the basis of the activity, and is working to analyze the measures, package them into groups that seem to make sense, come up with both micro and macro impacts on the economy and look at the impacts of the measures individually.

That is the stage that we are at right now. This group is off doing its work. It has produced some results, but it has not quite finished yet.

I will just refer to a couple of milestones. I have already mentioned the measures working group report, the 400-page catalogue. It includes nine so-called cross-cutting or foundation measures [along the lines of some of the ones that you will see in the U.S. plan] and 84 sectoral measures. And they are in the predictable kinds of sector groups: the residential, commercial, industrial, transportation and so on.

The kinds of measures that you will find in the catalogue cut across the range of things, whether they be voluntary or market based or regulatory, and there is probably a preponderance of voluntary type measures in this approach.

The Forecast Working Group, as I said, is analyzing the measures for micro and macro impacts. We have some results available. One of the things we have tried to do is package these measures in ways that seem to make some sense for both analytical purposes and for purposes of considering them in a program.

Modeling scenario one is a package of about 35-or-so measures from the catalogue that everybody agrees should happen. Building on that, modeling scenario two adds to the package another set of measures that are a little bit more intrusive, a little bit more challenging to implement, have a bit less acceptance and might be a slightly more difficult to implement. And so on up the scale.

Items in modeling scenario three are even more difficult to implement.

Modeling scenario four is the package that begins to jack up the intensity of specific measures within one, two and three. And it is within this range that we expect to find the stabilization goal. So in other words, we have discovered that we probably need most of the measures to get there.

Modeling scenario five has a lot of question marks in it, these are ideas that haven't jelled yet.

There is lot of information to come in terms of what the impacts on the economy will be, what the employment implications are, and what government spending implications would be for implementation.

We have some ideas as to the private investment costs so far and about the energy savings and we have a handle on what the expected impact would be in terms of tons of CO_2.

We summarized all of this information in a September 15th report which is called "Report on Options for Canada's National Action Program on Climate Change." This report briefly summarizes each of the measures and provides information on the kind of thing that I have been providing here, as well as a bit of context setting at the front, a little bit of vision, a little bit of principle setting and a really sketchy outline of what a National Action Program could be.

This report is available and was used in a series of five meetings that were held across Canada during the first weeks of October in five cities, to get input and more ideas and buy-in.

Let me highlight some things that are proceeding on a fast track basis as we are developing the program. We have our own voluntary Climate Change Challenge Program. Ministers sent a lead letter to chief executive officers of various organizations fairly recently, challenging them to start getting their voluntary measures together, stating that this would be an important part of the National Action Program. These responses are starting to flow back now, and this is going to be an important part of the plan.

There is also a process underway to develop a registry for voluntary actions, something that would also be useful in an international context as it relates to joint implementation projects.

By the way, Canada is not waiting for this whole process to come to closure before we take action, there are other things that are happening as well as we move ahead.

For example, recently Canada announced a new FBI. This has a different connotations than in the U.S. In Canada it is called the Federal Buildings Initiative, a retrofit for federal buildings, and it involves energy service companies in doing the kinds of things that I am sure you are quite familiar with.

Also, there is a multi-stake holder task force which has been active, formed by the Ministers for Finance and Environment, that is looking at a range of fiscal opportunities that might be there that could be included in Canada's next budget. It is a barriers and disincentives kind of approach where we are looking for ways to take fiscal steps.

Several of the measures, by the way, that the task force is considering have emerged from the Climate Change Plan process and are now being considered in the budget making process.

So, the next steps. The development and finalizing of the National Action Program has moved from this large multi-stake holder process to a funneling down, and a writing team made up of people from the Federal and Provincial governments — a small writing team with just a half a dozen people or so — is hard at work. We are expecting to have a February 1995 meeting of the same Ministers to approve the National Action Program.

We also know that we are going to want to renew this multi-stake holder process. The group, the Climate Change Task Group that has developed the program to this point has been thanked for their efforts, and we are expecting that there will be a need to create another national multi-stake holder process that might have a different emphasis. It might be shifting from program development to program implementation.

We expect to present Canada's finished program at the First Conference of the Parties to the Framework Convention at the 1995 meeting in Berlin.

27

CLIMATE PROTECTION IN GERMANY

Wolfram Schoett

German Embassy
4645 Reservoir Rd., NW
Washington, DC 20007

On December 9, 1993, the Federal Republic of Germany submitted its ratification declaration on the Framework Convention on Climate Change to the United Nations. Already in August 1993, the German government had presented a preliminary report entitled "Climate Protection in Germany" in anticipation of Article 12 of the Convention. This paper is based on the first full report published in September 1994, which is an updated, substantially revised and expanded version of that preliminary report.

Table 1 provides a summary of emissions of the most important greenhouse gases (with the exception of chlorofluorocarbons (CFCs), chlorocarbons (CCS) and halons) in Germany in 1990. The "IPCC Draft Guidelines for National Greenhouse Gas Inventories" were used.

The emissions data for carbon dioxide (CO_2), nitrogen oxides (NOx), and carbon monoxide (CO) can be considered well-founded, as can the data for NMVOC, with some exceptions, while the data for CH_4 and, especially, for N_2O, must be considered subject to major uncertainties. Figure 1 shows the development over time, from 1970 to 1993, of CO_2 emissions, pursuant to data from the Federal Environment Agency; Figure 2 shows the development of CH_4 emissions from 1970 to 1992. The database that would permit a similar representation for N_2O is lacking.

CO_2 emissions in Germany (not including high-seas bunkering and international air transport) decreased from 1,068 million tonnes in 1987 (the reference year for the Federal Government's CO_2-reduction resolution) to 911 million tonnes in 1993. This corresponds to a 14.7% reduction within this period. In the area of the former GDR, CO_2 emissions decreased by nearly 50% during this period (the main reasons for this decrease were economic restructuring, a reduction in population size by about 6%, a partial transfer of production to former West Germany, improvements in energy-use efficiency and a decrease in consumption of lignite, which is an intensive source of CO_2 emissions). CO_2 emissions in former West Germany were some 2% higher in 1993 that they were in 1987, but the population in the area of former West Germany also grew by some 7% from 1987 to 1993.

Methane emissions in Germany decreased by some 12% between 1970 and 1992, to 6,200 kilotonnes per year. In contrast with the decreasing emissions in former West Germany, emissions in the area of the former GDR increased until 1989. This trend was reversed in 1990, however, through a drastic reduction in livestock inventories.

Evaluating Climate Change Action Plans, edited by White
Plenum Press, New York, 1996

Table 1.

Sources and sinks of greenhouse gases	CO_2 Mt/a	CH_4 kt/a	N_2O kt/a	NO_x as NO_2 kt/a	CO kt/a	NMVOC not incl. CFCs/ CCs kt/a
Total emissions	**1012**	**6218**	**223**	**2944**	**10768**	**2978**
1. Energy-related	**983**	1767	33	**2923**	**10104**	**1679**
2. Industrial processes	29	11	**100**	21	664	129
3. Use of solvents and products			6			1170
4. Agriculture		2043	80			
5. Land use changes and forestry	-20					
6. Waste management		**2397**	4			

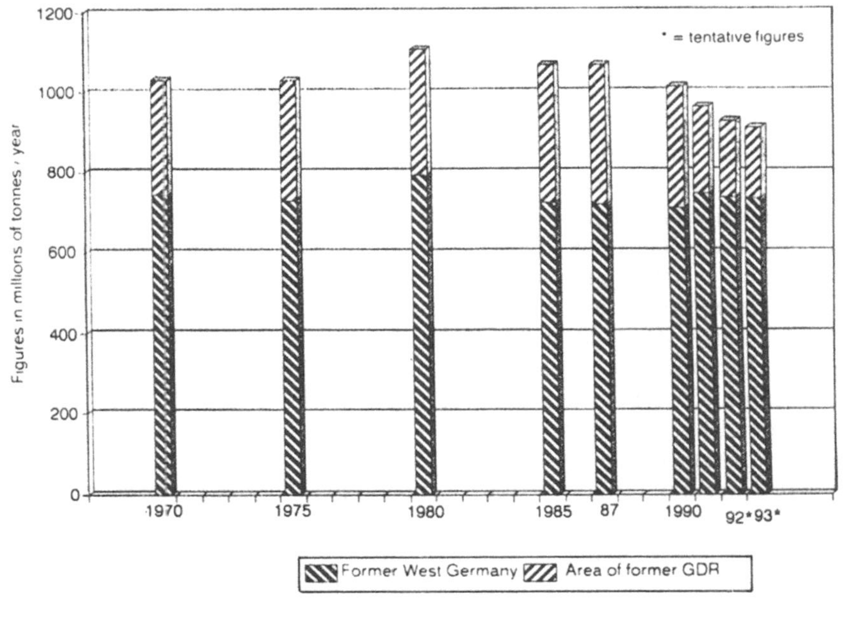

Figure 1. Development of CO_2 emissions in Germany 1970-1993.

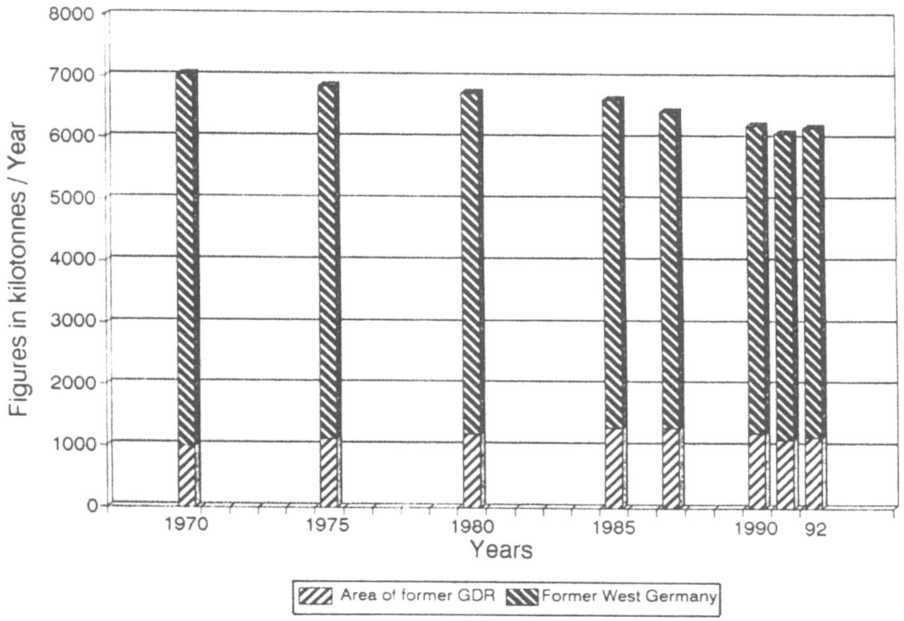

Figure 2. Development of CH_4 emissions in Germany 1970-1992.

The annual carbon-level increase in the existing 10.8 million hectares of forest is some 5.4 million tonnes (some 20 million tonnes of CO_2).

In Germany, it is mainly natural and near-natural ecosystems that seem threatened, when the country's geographic and climatic conditions are considered. Effects on water resource management and on agriculture and forestry, which are particularly sensitive to climatic influences, have a more direct influence on human living conditions.

The federal government considers measures for reducing emissions on climate-relevant gases to have priority. It is also emphasizing research into the consequences of climate change, however, in order to develop effective strategies for adaptation to the effects of such change, which cannot be ruled out, in spite of a wide range of efforts to reduce emissions of greenhouse gases.

By means of resolutions of June 13, 1990, November 7, 1990, and December 11, 1991, the Federal Cabinet approved a comprehensive CO_2 reduction program. Its aim in this connection is to reduce CO_2 emissions by 25% to 30% by the year 2005, in relation to the emissions volume in 1987. The Federal Government is aware of the difficulty in achieving this, a difficulty that is also due to the changed global framework. Reduction of other climate-relevant emissions, such as methane (CH_4), nitrous oxide (N_2O), (NO_x), (CO) and methane-free volatile organic compounds (NMVOC), is also taken into account in the national climate-protection strategy.

Overall, the Federal Government is aiming to achieve a reduction of all climate-relevant emissions – expressed in terms of CO_2 equivalents – of an order of 50% by the year 2005, in relation to 1987 levels.

The National Action Plan contains a very comprehensive catalogue of measures. On the one hand, this catalogue provides an overview of approved and implemented measures to date; on the other, it lists measures that are currently being approved by the appropriate decision-making bodies, or whose approval is currently being prepared or planned. It

Table 2A. Measures of the Federal Government contributing to the reduction of emissions of CO_2 and other greenhouse gases (109 measures in total) – Energy supply

	Measures
(1)	Federal table of charges for electricity
(2)	Support for local and regional energy-supply and climate-protection concepts
(3)	Act on the Sale of Electricity to the Grid
(4)	Elimination of the excise duty on lamps
(5)	Federal Government/Länder district-heating modernisation programme for the area of the former GDR
(6)	Funding for renewable energies
(7)	Energy-saving programme of the European Recovery Programme (ERP)
(8)	Support for advising of small and medium-sized companies concerning energy-saving
(9)	Support of the *Forum für Zukunftsenergien* (Future Energy Forum)
(10)	Information on use of renewable energies
(11)	Information on saving energy and efficient energy use
(12)	Acceleration of authorisation procedures by removing wind-energy systems from the 4th Ordinance on Execution of the Federal Immission Control Act (*4. BImSchV*)
(13)	Tax breaks for heat/power cogeneration plants

	Measures
(89)	Amendment to the Energy Management Act
(90)	Proposal for an Ordinance on Heat Use

comprises the areas of private households and small consumers, traffic and transport, industry, the energy industry, the waste-management industry, and agriculture and forestry.

Table 2 provides an overview of some 100 individual measures that are contributing to reduction of emissions of CO_2 and other greenhouse gases.

Research into climate systems, and into the consequences of climate change, is among the emphases of German environmental research. Climate-system research is seeking to obtain reliable statements concerning the development of climate change and, especially, concerning the development of regional climate. The purpose of research into the consequences of climate change is to estimate the possible effects of climate change.

The Federal Government is also sponsoring research into means of easing the effects of climate change. These efforts are aimed at providing the necessary action-oriented knowledge, and technology, in the areas of environmental protection and energy, for reduction of greenhouse gas emissions. They are also focused on identifying options for action to deal with the effects of climate change. Table 2 (especially in its "New Technologies" section) lists specific measures for supporting research into means of easing the effects of climate change.

Because global climate change is a long-term process, education, training and promotion of public awareness are of central importance. This is why the Federal Government and the states are conducting a campaign of intensive environmental information. The

Table 2B. Traffic and transport

	Measures
(14)	Increase of the mineral-oil tax
(15)	Emissions-oriented motor-vehicle tax (1st phase)
(16)	1992 Federal Traffic Infrastructure Plan
(17)	Increasing the attractiveness of local public transportation
(18)	Lowering of NMVOC emissions from vehicle refuelling, as a result of the Gas-balance System Ordinance
(19)	Programme of research into urban traffic (*Forschungsprogramm Stadtverkehr - FOPS*)
(20)	Improving continuity of traffic flow
(21)	Information on energy-saving and environmentally aware driving habits
(22)	Research projects and information concerning urban traffic planning and decreasing traffic pollution in cities
(23)	Railway structural reform
(24)	Freight centres
(25)	Combined transport using waterways
(26)	Research programme on "Pollution in Aviation"
(27)	Traffic research
(28)	Deregulation of goods transports on roads by means of the Tariffs-elimination Act (*Tarifaufhebungsgesetz*)

	Measures
(91)	Raising of the minimum EU levels for mineral-oil tax
(92)	Emissions-related motor-vehicle tax (2nd phase)
(93)	Tolls for use of certain roads
(94)	CO_2 emissions of new motor vehicles
(95)	German railways' site concept
(96)	Use of modern information technology for preventing and regulating traffic (telematics)
(97)	Taxation of aircraft fuels
(98)	Amendment of the Common Rules of Procedure (*Gemeinsame Geschäftsordnung*) of the federal ministries
(99)	Introduction of traffic-effects studies
(100)	Shifting of international transit traffic from roads to railways and to ships

Table 2C. Buildings and structure

	Measures
(29)	Amendment of the Thermal Insulation Ordinance (*Wärmeschutzverordnung - WSchV*)
(30)	Amendment of the Heating-systems Ordinance (*HeizAnlV*)
(31)	Advising concerning energy-saving, efficient energy use in housing structures - on-site advising
(32)	Tax breaks through the Support-area Act (Fördergebietsgesetz), pursuant to the 1991 Tax-Amendment Act of 24 June, 1991 and the Act on Securing the Futures of Sites (*Standortsicherungsgesetz*) of 13 September, 1993
(33)	Housing modernisation programme of the Kreditanstalt für Wiederaufbau (KfW) reconstruction bank
(34)	The joint programme "Economic Recovery in eastern Germany"
(35)	Subsidies for construction of public (low-rent) housing
(36)	Programme of research into experimental city and housing planning: "Reducing pollution in city planning" (*Schadstoffminderung im Städtebau*)
(37)	Acceleration of authorisation procedures by means of the investment-facilitation and housing-construction-site act (*Investitionserleichterungs- und Wohnbaulandgesetz*)
(38)	Reduction of barriers to investment in housing construction in the area of the former GDR, for cases in which ownership questions have not been settled
(39)	Information for building owners, architects, planners, engineers, craftsmen

	Measures
(101)	2nd Ordinance for Amendment of the Ordinance on Small Combustion Plants (1st Federal Immission Control Ordinance)
(102)	Instruments for increasing energy efficiency of existing buildings
(103)	Special relevant privileges accorded within the building code
(104)	Standardisation of authorisation practices for renewable energy systems

Table 2D. New technologies

Measures	
(40)	Specialised programme on environmental research and technology
(41)	Research into, and technical refinement of, power-plant and firing-plant technology, especially clean coal-firing technology
(42)	Research and development concerning gas and steam turbine power plants
(43)	Research and development concerning use of renewable energies
(44)	Programme for promotion of photovoltaic systems
(45)	Support for testing wind-power systems: "250 MW Wind"
(46)	"Solarthermie 2000" solar-energy promotion programme
(47)	Research and development concerning use of solar energy
(48)	Research and development concerning secondary energy systems integrated with renewable-energy systems
(49)	Research and development concerning efficient energy use
(50)	Nuclear energy research/reactor-safety research
(51)	Nuclear-fusion research
(52)	Research into thermal treatment of waste
(53)	Test project entitled "Generation of heat and power from renewable raw materials"
(54)	Geothermic energy

Table 2E. Agriculture and forestry

Measures	
(55)	Joint task "Improvement of Agricultural Structure and Coastal Protection"
(56)	Bonuses for land set-asides
(57)	Improvement of animal digestive efficiency as part of animal husbandry, in order to reduce methane emissions
(58)	Support for extensive methods of agriculture
(59)	Conservation of existing forests
(60)	Support for new afforestation
(61)	Forest-management measures
(62)	Tax exemptions for pure rape methyl ester (RME)
(63)	Agency for renewable raw materials (*Fachagentur Nachwachsende Rohstoffe*)
(64)	Proposal for a fertilizer ordinance

Table 2F. Waste-management sector

Measures	
(65)	Ordinance on packaging
(66)	Technical Instructions on Municipal Waste Management (*TA-Siedlungsabfall*)
(67)	Technical Instructions on Waste, Part 1
(68)	Waste Management and Product Recycling Act

Table 2G. Overarching measures

	Measures
(69)	Improvement of training and continuing training for architects, engineers, technicians and craftsmen
(70)	Support programme of the Deutsche Bundesstiftung Umwelt (DBU)
(71)	Investment programme for reducing environmental pollution (Federal Ministry for the Environment, Nature Conservation and Nuclear Safety)
(72)	Environmental programme of the Kreditanstalt für Wiederaufbau (KfW) reconstruction bank
(73)	Environmental programme of the Deutsche Ausgleichsbank
(74)	Environmental protection guarantee programme: liability exemption in connection with the supplementary loans programme III (*Ergänzungsdarlehen III*) for subsidising manufacturers of preventative environmental protection technology
(75)	Federal/Länder joint task "Improvement of the regional economic structure"
(76)	Financial support of the economic infrastructure in the area of the former GDR - improvement of the regional economic structure in connection with support for municipal infrastructures
(77)	Advising concerning thrifty, efficient energy use, provided by the *Arbeitsgemeinschaft der Verbraucherverbände (AgV)* consumer associations' working group, on commission to the Federal Ministry of Economics
(78)	Support for advising of small and medium-sized companies concerning environmental protection and energy use
(79)	Orientation advising on environmental protection for small and medium-sized companies (area of the former GDR)
(80)	Orientation advising on environmental protection for communities in the area of the former GDR
(81)	Community loan programme - area of the former GDR
(82)	Air quality control programme of the European Recovery Programme (ERP)
(83)	Technical information concerning efficient energy use and use of renewable energies
(84)	Studies on optimising the CO_2-reduction programme
(85)	Amendment of the Fee Table for Architects and Engineers (HOAI)
(86)	Research into specific approaches to regulatory tools
(87)	System analysis within the Instruments for Climate-gas-reduction Strategies (IKARUS) project
(88)	Environmental certification mark (*Umweltzeichen*)

(105)	Improvement of the framework for vocational training and for continuing education and training
(106)	Support for provision of information concerning third-party financing models
(107)	Introduction of an (at-least) EU-wide CO_2/energy tax that has a neutral effect with regard to competition and total tax-revenue volume
(108)	Law on mandatory labelling with regard to energy consumption (product labelling)
(109)	Planning of the parliament and government district in Berlin with regard to environmental protection requirements, especially climate-protection requirements

Table 3. Emissions scenario

	Emissions change in %	
	1987 - 2005	1990 - 2005
Methane	-50	-48
Nitrous oxide	n.a.	-25
Nitrogen oxide	-36	-25
Carbon monoxide	-58	-51
NMVOC	-47	-43

topic of environmental protection has been solidly integrated into school education, which lies within the responsibility of the Länder (states).

The Federal Government alone will not be able to locally implement such an intensive program that affects economic and social structures. Gradually, the Länder (states) are also preparing their own Land/(state-)specific climate-protection programs. Since 1990, many communities have begun developing and implementing community CO_2-reduction concepts, often on the basis of existing energy-supply concepts. To date, far more than 100 such concepts have been developed.

The development of energy-related and non-energy-related emissions of methane (CH_4), nitrous oxide (N_2O), nitrogen oxides (NO_x), carbon monoxide (CO) and methane-free volatile organic compounds (NMVOC) until the year 2005 was estimated, for the core of energy-related emissions on the basis of two existing studies, for the non-energy-related emissions on the basis of the current legal framework.

Table 3 lists the emissions reduction for these greenhouse gases (both energy-related and non-energy-related) until the year 2005, in relation to 1987 (where data are available) and 1990 emission levels.

In the area of energy, a focus of the Federal Government's cooperation in the international area of development, some 13 billion DM were provided between 1961 and 1993, within the framework of financial cooperation; some 1.1 billion DM were provided within the framework of technological cooperation. Of this funding, some 2.5 billion DM were spent on hydroelectric power generation; some 500 million DM were spent on other renewable energies. In addition, the Federal Government spends 300 million DM annually in helping other countries conserve their tropical rain forests.

In a pilot phase lasting from 1991 to 1993, and which was supported with approximately 1 billion DM worth of special-draft loans (central fund plus co-financing), the Federal Republic of Germany paid $147 million US into the central fund of the Global Environmental Facility (GEF). For the period lasting from mid-1994 to mid-1997, the Federal Republic of Germany has committed itself to payment of $240 million US (ca. 12% of the total volume of some $2 billion US) into this fund.

The Federal Government will continue to take an active role in implementing national and EU-wide climate-protection strategies, and in efforts to achieve a globally coordinated strategy for combatting the anthropogenic greenhouse effect. This applies especially to implementation and promotion of further development of the Framework Convention on Climate Change. The 1st Conference of the Parties, to be held in Berlin from April 7 to May 28, 1995, will be an important step in these efforts.

28

SESSION IV DISCUSSION WITH RICHARD FINDLAY AND WOLFRAM SCHOETT

Mr. Jhirad: We were very impressed by the inventories that have been put together and their evolution. I think the German figures are very ambitious and very impressive, and especially the discussion of some policies instruments. I think to an American audience, $10 per gallon of gasoline is absolutely inconceivable. I was shocked at having to pay $50 last month to fill up a 10-gallon tank when I was traveling in Germany.

An annual increase in fleet fuel efficiency of three percent a year in Canada between the years 1996 and 2000, following the U.S. lead is projected. Is this realistic?

Mr. Findlay: First of all, the motor vehicle market in Canada is very much a part of the North American motor vehicle market, and the measures that Canada can take alone are very limited. I can't comment on the number itself and whether that's particularly realistic.

Mr. Jhirad: Given Canada's cross-cutting air issues approach, how might climate change programs and studies affect ongoing U.S.-Canada negotiations on acidic precipitation?

Mr. Findlay: We realize that Canada has a large challenge in terms of reducing acidic precipitation, particularly in the eastern portion of Canada where quite a large percentage of the acidic precipitation is caused by emissions in the United States. We recognize that steps taken to eliminate emissions in the climate change area will probably have positive effects on the acid rain site as well.

At the moment, we are looking at benefits that cut across those two issues and I don't think I can really give a more definitive answer.

Mr. Jhirad: Why did Germany choose not to report on the year 2000, as indicated in the INC instructions?

Dr. Schoett: In Germany the discussion about the risks of carbon dioxide emissions started before the Rio conference. In 1990 we decided to set reduction goals of 25% of the 1987 emissions levels by the year 2005. The decision was based on the recommendations of the Enquete Commission "Protecting the Earth's Atmosphere," established in 1987. It's a very ambitious goal and more than is demanded by the Convention for the year 2000. Any figure less than 25% would have been regarded as a step back and could have diminished the pressure on industry to take appropriate measures.

The Enquete Commission also recommended stopping the production and use of the most dangerous CFCs in Germany by the year 1995, far beyond the Montreal Protocol.

Mr. Jhirad: Given the strong position or posture of Germany in reducing carbon emissions, first, why does Germany subsidize the use of coal? And two, why did Germany take 18 months to ratify the convention. Germany was the 45th country to ratify?

Dr. Schoett: The problem of coal subsidizing has been under debate for decades. The problem is we have a large and traditional mining industry in the Ruhrgebiet and in other parts of the country and many people are dependent on the mining of coal. Although the government has tried to reduce, and they have reduced, subsidies for coal mining, it is still forced to continue subsidizing for social reasons, in order not to give up hundreds of thousands of jobs. We are paying a high price since German coal is not at all competitive on the world market — it's price is two to three times the world market price.

It took Germany 18 months to ratify the Convention because any agreement between governments has to pass our two chamber system, which takes time. However, months before ratification by the parliament, Germany published a preliminary report on climate protection, one year earlier than asked for by the convention. There never was any doubt that the Convention would pass parliament.

Mr Jhirad: You indicate that emissions in West Germany in '93 are higher than in 1990. Can Germany stabilize at 1990 levels by the year 2000, given that rise?

Dr. Schoett: There has been a small rise in the past few years, but this rise is explained by an increase of population in West Germany and by the transfer of industries from the eastern part of Germany to West Germany.

The government is still convinced that they can achieve the goal of 25% by the year 2005 which is more ambitious than stabilizing 1990 levels by the year 2000.

Mr. Jhirad: Are there any cost estimates or assessment of effectiveness of the German mitigation measures? In looking at these various options, voluntary, market-based or regulatory, you mentioned the taxes, do you have any estimates of the cost effectiveness of these relative policy instruments?

Dr. Schoett: Well, it's difficult for me to answer this question, because I have not worked in this field. As a matter of fact,if Germany is raising taxes on energy consumption, these taxes will not serve as additional income to the government but rather be counterbalanced by tax reductions in other fields. This should give an incentive to the public to save energy.

If one talks about a CO_2 tax, an issue which we will present to the European Union, it is clear that some industries, such as energy demanding branches like the aluminum industry, will suffer most. Therefore our industrial associations argue strongly against the idea of an energy/CO_2 tax.

Mr. Jhirad: Mr. Findlay, would you like to address that?

Mr. Findlay: We do have some information available on the cost effectiveness of measures and packages of measures, but it's not complete. What we are striving to do is put together information that reflects both costs and benefits and does so in a broad sense; sectoral, regional, and employment sense.

Mr. Jhirad: Mr. Findlay alluded to Canada's need to get provincial input into these climate action plans, and the issue of provincial input is quite important, because Agenda 21, the Framework Convention and other Rio initiatives say that climate change activities should solicit input from sub-national groupings in the country, whether these are states or provinces.

So the real question is how important is this input from these sub-national entities? What are the barriers and how, if at all, are states and provinces involved? How is that input to come from the bottom up to the federal level?

29

JAPAN'S ACTION REPORT ON CLIMATE CHANGE

as presented by Satoshi Tanaka

Embassy of Japan
2520 Massachusetts Ave. NW
Washington, DC 20008

I. JAPANESE CONTEXT

Japan is an archipelago stretching approximately from 24 degrees to 46 degrees north latitude. Its territory as of 1990 extends over 37,770,000 hectares, or about 0.3 percent of the earth's land area. Japan's population of about 123 million is 2.5 percent the world population. Its gross domestic product was about 433 trillion Yen ($2,958 billion US) in 1990. About 67 percent of Japan's land is covered by forest. Japan's climatic zones range from subtropical to subarctic, with four sharply distinct seasons. The long-term rate of increase in Japan's average annual temperature is currently estimated to be 0.9 degrees Celsius per 100 years.

Japan is dependent on foreign sources for most of the energy that produces the greater part of its carbon dioxide emissions. Though this dependence has been reduced somewhat by introducing oil substitutes in recent years, it remains slightly above 80 percent, leaving Japan highly exposed to the effects of supply fluctuations. Final energy consumption increased substantially throughout the period of rapid economic growth of the 1960s (during which real annual economic growth averaged 10.3 percent), but since the first oil crisis energy consumption has tended to remain level or to sink: the ratio of unit energy to unit GNP in fiscal 1990 was 36 percent less than in fiscal 1973. Japan's per capita energy consumption is therefore extremely low by advanced industrialized nations' standards, the equivalent of about 4,250 liters of oil per annum. Broken down, energy consumption was increasing considerably in the industrial, commercial/residential, and transport sectors of the economy until the first oil crisis; from 1973 until 1986, consumption tended to continue to increase in the commercial/residential and transport sectors, whereas consumption levels generally continued to decline in Japanese industry, which had accomplished a structural conversion to low energy consumption and the world's most advanced high-energy-efficiency technology. All three sectors tended to increase energy consumption as long as the economy remained strong after 1986; energy consumption continued to grow especially in the commercial/residential and transport sectors even after the economy began an adjustment phase in mid 1991, but in industry, it declined. In short, although total emissions have

Evaluating Climate Change Action Plans, edited by White
Plenum Press, New York, 1996

continued to increase over the past few years, per capita carbon dioxide emissions are lower, thanks to Japan's generally temperate climate, relative geographical exiguity, and advanced economy, in addition to efforts to reduce energy consumption since the first oil crisis and greater resulting energy efficiency.

II. NATIONAL INVENTORY OF GREENHOUSE GAS EMISSIONS AND REMOVALS

1. Basic Approach

This greenhouse gas emissions and removals inventory was compiled in accordance with the following approach based on the Guidelines for the Preparation of First Communications by Annex I Parties.

An inventory was compiled for emissions of carbon dioxide (CO_2), methane (CH_4), nitrous oxide (N_2O); and precursors: other nitrogen oxides (NO_x), carbon monoxide (CO), and non-methane volatile organic compounds (NMVOC) during fiscal 1990 (April 1990-March 1991). Removals of carbon dioxide were also calculated.

Total greenhouse gas emissions and removals were calculated generally by the methods explained in the IPCC/OECD Draft Guidelines. That is, to quantify greenhouse gas emissions and removals in each sector, activity data on fuel consumption or other relevant values were multiplied by an emission factor for each gas emitting source and a removal factor for each sink. Significant figures were considered in the calculations. To calculate nitrogen oxide emissions, total measured nitrogen oxide emissions from all soot and smoke emitting facilities controlled under the Air Pollution Control Law were summed.

Currently, emission/removal factors and activity data needed to estimate emissions and removals are insufficient in certain categories. Improvement in these fields will be possible as more information is acquired and as international trends develop.

2. National Inventory of Greenhouse Gas Emissions and Removals

Table 1 contains Japan's fiscal 1990 greenhouse gas emissions and removals inventory.

III. POLICIES AND MEASURES

In October 1990 the Japanese government established an Action Program To Arrest Global Warming by decision of the Council of Ministers for Global Environment Conservation. This Action Program spells out the governments current orientation in the short term and the general framework of enforceable measures that it should take in the future to promote planned and comprehensive measures against global warming; it also clarifies its basic approach toward obtaining the Japanese peoples understanding and cooperation as well as contributions within the international framework. The Action Program is the national program of Japan provided for in Article 4 paragraph 1(b) of the United Nations Framework Convention on Climate Change.

The Action Program defines the formation of an environmentally sound society, the harmonization of environmental protection with stable economic development, and international coordination as "basic elements" of efforts to counteract global warming; it sets the year 2000 as its interim target year and the period until 2010 as its duration. Its targets are defined below:

Table 1. National greenhouse gas inventory in fiscal 1990

(unit: Gg)

Greenhouse gas source and sink categories	CO_2	CH_4	N_2O	NO_x	CO	NMVOC
Quality (confidence in estimation)	High	Medium	Low	High	Low	Low
Total National Emissions	1,173,000	1,380	48	1,898	2,809	2,060
I. All Energy (Fuel Combustion + Fugitive)	1,075,000	125	22	1,844	2,792	560
A. Fuel Combustion	1,075,000	25	22	1,844	2,792	340
Energy & Transformation Industries	82,000	2	5	388	126	40
Industry (including Agriculture/Forestry)	489,000	8	4	393	306	20
Commercial/Institutional	123,000	0.5	NE	16	5	NE
Residential	139,000	0.5	0.1	38	25	NE
Transport	215,000	14	13	1,009	2,330	280
Other	9,000	NE	NE	NE	NE	NE
Biomass Burned for Energy	18,000	NE	NE	NE	NE	NE
B. Fugitive Fuel Emissions	NE	100	NE	NE	NE	220
Oil and Natural Gas Systems	NE	NE	NE	NE	NE	220
Coal Mining	NA	100	NA	NA	NA	NE
II. Industrial Processes	53,000	NE	15	1	NE	60
A. Chemicals	NE	NE	15	1	NE	60
B. Non-metallic Mineral Products	43,200	NE	NE	NE	NE	NE
C. Other	9,800	NE	NE	NE	NE	NE
III. Solvent Use	NE	NE	NE	NE	NE	1,440
IV. Agriculture	NE	790	5	NE	NE	NE
A. Enteric Fermentation	NA	330	NA	NA	NA	NA
B. Animal Wastes	NA	190	NA	NA	NA	NA
C. Rice Cultivation	NA	261	NE	NA	NA	NA
D. Agricultural Soils	NA	NE	4	NA	NA	NA
E. Agricultural Waste Burning	NA	6	1	NE	NE	NE
F. Savanna Burning	NA	NA	NA	NA	NA	NA
V. Land Use Change & Forestry	IE*	NE	NE	NE	NE	NE
A. Forest Clearing & On-Site Burning of Cleared Forests	NE	NE	NE	NE	NE	NA
B. Grassland Conversion	NE	NA	NA	NA	NA	NA
C. Managed Forests	IE*	NA	NA	NA	NA	NA
D. Abandonment of Managed Lands	NE	NA	NA	NA	NA	NA
VI. Waste	45,000	465	6	53	17	NE
A. Landfills	1,000	446	NA	NA	NA	NA
B. Wastewater	NA	6	NE	NA	NA	NA
C. Other	44,000	13	6	53	17	NE
International Bunker Oil	31,000	NE	NE	NE	NE	NE
Total Removals (Managed Forests)	90,000	NA	NA	NA	NA	NA

Code Meaning
NA = Not applicable
NE = Not estimated
IE = Estimated but included elsewhere
* = Estimated taking into account removals in sink data

- The Government of Japan, based on the common efforts of the major industrialized countries to limit CO_2 emissions, establishes the following target for the stabilization of Japan's CO2 emissions.
  - The emissions of CO_2 should be stabilized on a per capita basis in the year 2000 and beyond at about the same level as in 1990, by steadily implementing a wide range of measures under this Action Program, as they become feasible, through the utmost efforts by both the government and private sectors.
  - Efforts should also be made, along with the measures above, to stabilize the total amount of CO_2 emissions in the year 2000 and beyond at about the same level as in 1990, through progress in the development of innovative technologies, etc.,

including those related to solar, hydrogen and other new energies as well as fixation of CO_2 at the pace and in the scale greater than currently predicted.

- The emission of methane gas should not exceed the present level. To the extent possible, nitrous oxide and other greenhouse gases should not be increased.

With respect to sinks of CO_2, efforts should be made to work for the conservation and development of forests, greenery in urban areas and so forth in Japan and also to take steps to conserve and expand forests on a global scale, among others.

The Action Program also puts the following measures on the government agenda: measures to limit CO_2 emissions; measures to reduce the emissions of methane and other greenhouse gases; measures to enhance carbon dioxide sinks; promotion of scientific research and observation/monitoring; the development and dissemination of technology and the promotion of public awareness; and international cooperation..

The Action Program's implementing framework consists of annual followup by the Council of Ministers for Global Environment Conservation to review progress toward implementation and the latest data on carbon dioxide emission levels. The Council of Ministers has also decided to issue an appeal to the world community regarding the need for planning of long term vision to meet global warming ("The New Earth 21"). In November 1993 the Japanese Diet enacted the Basic Environment Law Japan's new law on the environment establishes as basic principles first, the enjoyment and perpetuation of the blessings of a healthy environment, second, the creation of a society ensuring sustainable development with reduced environmental load, and third, active promotion of global environmental conservation through international cooperation. It establishes a Basic Environment Plan in which the Action Program To Arrest Global Warming will be appropriately incorporated, it specifies measures that the national government must take, including surveys and research related to economic measures, and it provides for international cooperation for global environmental protection.

1. Measures to Limit Carbon Dioxide Emissions

1.1. Industry. In industry, which accounts For about half of Japans carbon dioxide emissions, the development of energy-saving technology has been promoted under the Energy Conservation Law since the first oil crisis; measures have been taken to improve and better apply standards related to the rationalization of fuel combustion and to assist investments in energy-saving capital equipment through special taxation measures and low-interest financing.

Through these measures, energy use has become more efficient, and marked improvements have been made in energy consumption per mining and industrial production unit. The major energy-saving investments are now already productive; the Energy Conservation Law was amended to effect further energy savings, and the Energy Conservation and Recycling Assistance Law, enacted in 1993, has radically strengthened policies by providing very low interest financing and other incentives for related capital investments.

Japan is also studying energy conservation measures aimed at the use of types of energy yet untried in agriculture, forestry and fisheries, and construction. The government is also urging manufacturers to set their own objectives for reducing carbon dioxide emissions.

1.2. Residential and Commercial. Energy consumption is tending to rise in both the residential and the commercial/institutional sectors, due to advances in office automation and the wider use and growing capacity of electric home appliances. The following measures are being taken in the residential/commercial sector. i) standards related to insulation in construction are being improved and more strictly applied under the Energy Conservation Law and steps are being taken to assist builders through extra financing, etc., ii) Energy Conservation Law standards for home appliances, etc., are being strengthened and their scope expanded, iii) the utilization of waste heat and other yet-unused forms of energy is being promoted through low-interest financing, etc., iv) demand for energy for cooling is being curbed through measures to mitigate the heat island phenomenon by planting more greenery in urban areas, and v) the planning of cities that emit little carbon dioxide is being promoted using subsidies from the national treasury and other means.

1.3. Transport. Motor vehicles account for a large and growing proportion of energy consumed to transport both passengers and freight. The following measure are therefore being directed at the trans port sector. i) carbon dioxide emissions from motor vehicles are being curbed by setting and strengthening standards (as of 1993) related to gasoline engine passenger cars under the Energy Conservation Law and by using national treasury subsidies, special taxation measure, etc., to promote the introduction of low-emission vehicles, ii) the efficiency of freight transport is being increased by improving services and promoting the use of railway and coastal shipping using interest-free loans, special taxation measures, etc., and by promoting consolidated cargo transportation for intracity distribution, iii) the use of public means of transportation in passenger traffic is being promoted by increasing railway transport capacity and stimulating bus transportation, etc., and iv) transportation systems that generate less carbon dioxide are being created by improving transportation infrastructure to facilitate motor vehicle traffic based on the 11th Five-Year Road Improvement Plan (1993) and the 5th Specific National Five-Year-Project for Traffic Management Systems Installation (1991).

1.4 Energy Conversion. In order to use energy more efficiently and to build an energy supply structure that generates little carbon dioxide, Japan is promoting the development of technology (for solar and other new and recycled forms of energy and fuel cells and other new supply systems based on the Basic Plan for Energy Research and Development); and intending to improve the efficiency of power generation by thermal power plants and to promote energy sources such as nuclear power, LNG, and hydropower that generate less or no carbon dioxide. Japan is taking measures to support development in these fields through low-interest financing and special taxation measures.

1.5. Crossover Sector. As measures to realize an environmentally protective lifestyle that overlap multiple sectors, Japan is setting goals, standards, etc., by Recycling Law, promoting recycling through various supportive measures, increasing awareness of environmental protection through the use of environmental marks, and promoting the use of more appropriate packaging.

2. Measures to Enhance Carbon Dioxide Sinks (Changes in Land Use and Forests)

Japan has long had a large percentage of forested land and has maintained this high percentage. Difficulties stemming from depressed wood prices, increased wood-product imports, etc. in the forest and wood industries are, however, visibly exerting effects on forestry management in some areas. Japan is therefore seeking i) to establish Forest Plans under the Forest Law and to promote planned forest conservation and improvement through the adequate management of Protection Forests, afforestation, and the thinning of forests, ii) to promote the effective use of wood resources, iii) to suitably protect forest that is a foundation for Japan's natural environment through such measures as the designation of nature conservation areas, and iv) to conserve and manage greenery in urban areas through a variety of beautification programs.

3. Measures to Reduce Methane Emissions

3.1. Waste Management. About one-third of Japan's methane emissions is generated by waste landfills. Intermediate treatment and recycling are now reducing the volume of waste ultimately in landfills, but waste emissions are on the increase. The following measures are therefore being promoted to improve waste management: i) the Waste Treatment law was amended in 1991 to encourage people to reduce and recycle waste, ii) treatment facilities are being constructed with the aim of reducing landfill volume, and iii) treatment plants are being converted to types that generate less methane. Japan is also taking steps to curb the consumption of fossil fuels though more efficient use of power generated from waste and other uses of heat from waste incineration.

3.2. Agriculture. Measures to reduce methane emissions in agriculture are still at the research stage; studies arc underway to find ways to reduce the methane generated by rice cultivation and by livestock through enteric fermentation.

3.3. Energy Supply and Other Sectors. Efforts are being made to use the gas obtained during coal mining as fuel and legally to prohibit leaks during natural gas development and fugitive gas from gas works.

4. Measures to Reduce Nitrous Oxide Emissions

Japan is conducting surveys and research and developing technology with the object of quantitatively determining the amounts of nitrous oxide generated and emitted in manufacturing, agriculture, and waste management and introducing measures to reduce emissions.

5. Measures to Reduce Emissions of Other Greenhouse Gases

Regulatory measures have been written into the Air Pollution Control Law to systematically deal with nitrogen oxides, carbon monoxide, and non-methane volatile organic carbons. The Japanese Government has established environmental standards for maximum levels of nitrogen oxides and carbon monoxide that should not be exceeded from the standpoint of protecting human health and has set concentration guidelines for the achievement of environmental standards for photochemical oxidants that in effect regulate levels of non-methane volatile organic carbons. Factory, workplace, and motor vehicle emission regulations are being enforced with regard to nitrogen oxide, and financing and special taxation measures are carried out to encourage the building of soot and smoke treatment facilities.

6. Promotion of Public Awareness

Japan has improved its environmental guidance in the Courses of Study (the Ministry of Education's official guidelines for teachers) in addition to promoting environmental protection, resource and energy conservation, and greening campaigns through public relations and the setting of campaign periods in newspapers and other media. In order to promote environmental protection activities by private organizations, a jointly funded government and private-sector Japan Fund for Global Environment was established in 1993.

Finally, "Guidelines for Environmentally Sound Corporate Practices" and other manuals and guidelines are being drafted and distributed.

7. Scientific Research, Observation, and Monitoring

Scientific research, observation and monitoring, and development of technology related to global environmental problems are being conducted in the framework of two plans: the annually drafted Comprehensive Promotion Program for Global Environment Research, Monitoring and Technology Development and a long-term plan, the Basic Plan for Research and Development Related to Earth Science and Technology (1990). The Comprehensive Global Environment Research Promotion Budget and others have been established to integrate and complement surveys and studies related to global environmental protection. In particular, Japan is promoting regionally extensive observation and monitoring, and surveys and research involving participation and tie-ins with international groups for planning of global environmental research, observation and monitoring.

8. Promotion of International Cooperation (Including Funding and Technology)

Harmony between the environment and development was stated to be one of the principles of ODA (Official Development Assistance) in the ODA Charter, adopted by the Cabinet in 1992. Based on the Prime Minister's announcement at the UNCED, Japan endeavors to significantly expand its ODA in the field of the environment to around 900 billion to 1 trillion Yen during the five-year period starting from fiscal 1992. The Basic Environment Law also stipulates that Japan will promote international cooperation and other efforts directed at global environmental protection.

Specifically, Japan's contributions include \$48.20 million US (as of March 31, 1994) to the GEF core fund, human and budgetary contributions to the IPCC, the holding since 1991 of regular regional seminars to integrate promotion in the Asian-Pacific region of measures to arrest global warming and other comprehensive support for efforts to prevent global warming, the establishment of centers to provide information about environmental protection technology and transfers of energy-saving technology, support for the conservation and afforestation of tropical rain forests for the creation of carbon dioxide sinks, international cooperation to efficiently promote environmental and energy technology development, subsidies for NGOs, and support of private-sector international cooperation through the Japan Fund for the Global Environment.

IV. PROJECTIONS OF THE EFFECTS OF GREENHOUSE GAS COUNTERMEASURES

The projections of the effect of greenhouse gas countermeasures described here target the fiscal year 2000 (April 2000 to March 2001). The greenhouse gases considered in these projections are carbon dioxide, methane, and nitrous oxide. The Projection of the effect of measures to enhance greenhouse gas sinks target carbon dioxide.

1. Carbon Dioxide Emissions

1.1. Projection and Evaluation of Carbon Dioxide Emissions in Fiscal 2000. Projections of "Energy (Fuel Combustion)" sector, which accounts for a large proportion of carbon dioxide emissions, are based on the energy supply and demand outlook for fiscal

Table 2. Carbon dioxide emissions in fiscal 2000

Category	Emissions
Energy	3.1×10^8 tons of carbon (1.1×10^6 Gg-CO_2)
Industrial Processes	0.1×10^8 tons of carbon (0.05×10^6 Gg-CO_2)
Waste	0.1×10^8 tons of carbon (0.05×10^6 Gg-CO_2)
Total	3.3×10^8 tons of carbon (1.2×10^6 Gg-CO_2)

2000 in the "Long-Term Energy Supply and Demand Outlook" published by the Advisory Committee for Energy. The "Outlook" is based on assumed economic growth rate and the price of oil (see the notes below) and assumptions that efforts by all concerned parties will ensure that all energy-conservation measures taken since fiscal 1990 and new energy-conservation measures foreseeable as of fiscal 1994 will make an expected contribution to lower carbon dioxide emissions.

Total carbon dioxide emissions in fiscal 2000 resulting from those existing and new energy-conservation measures taken into account in the Long-Term Energy Supply and Demand Outlook are estimated to be about 330 million tons of carbon (Table 2).

On a per capita basis, this indicates emissions of about 2.6 tons of carbon per year in fiscal 2000; compared with the level actually calculated in 1990 (2.59 tons of carbon per year), this means that the first target of the Action Program cited above is estimated to be achievable.

Yet greater efforts will be necessary, however, to achieve the Action Program's second target of maintaining total carbon dioxide emissions at the 1990 level, since total emissions are estimated to increase with respect to the fiscal 1990 total of 320 million tons of carbon.

Notes:

Economic growth rates: For fiscal 1991-1994: actual rates through 1993 and from the Economic Outlook and Basic Policy Stance (1994).
For fiscal 1995-2000: 3.59% per annum, the rate assumed in the current economic plan "The Five-Year Economic Plan-Sharing a Better Quality of Life around the Globe." (fiscal 1992-1996), and its exploitations.

Oil Price: US$20 per barrel in fiscal 2000 (equivalent to current oil price in real term).

1.2. Projected Effects of Measures. If we compare totals based on the "Long-Term Energy Supply and Demand Outlook" assuming full compliance with energy-conservation measures with totals assuming no measures had been taken, it is estimated that overall these measures in fiscal 2000 will have an effect equivalent to approximately 30 million tons of carbon (about 120,000 Gg-CO_2). Without these measures, emissions will increase by approximately 10 million tons of carbon in each of the three subcategories under consideration: industry, commercial/residential, and transport.

In industrial process, it is estimated that a reduction in carbon dioxide emissions equivalent to about 2 million tons of carbon (7,000 Gg-CO_2) with respect to the 1990 level is possible by calcining less limestone in cement manufacture and so forth.

As for carbon dioxide emissions from waste, two municipal-waste scenarios were compared. The first assumes that no waste reduction measures are implemented and that the incineration ratio remains at the current level. The second assumes that municipal waste is reduced by 30% and that the incineration ratio continues to increase in line with past trends. Carbon dioxide emissions would be reduced by some 2 million tons of carbon (about 9,000 Gg-CO_2) in the second scenario as compared with the first.

1.3. Projected Levels beyond 2000. Total carbon dioxide emissions are expected to stabilize at the 1990 level beyond the year 2000 through implementation of medium- and long-term energy measures. Japan intends to maintain measures in step with world opinion to curb carbon dioxide emissions.

2. Carbon Dioxide Removals

Projected carbon dioxide removals in fiscal 2000 based on the forest management goals in the Nationwide Forest Plan (formulated every few years under the Forest Law) are approximately 25 million tons of carbon (about 92,000 Gg-CO_2), slightly better than in fiscal 1990.

It is necessary to continue to strive for better forest management to achieve the target of the Action Program To Arrest Global Warming: "With respect to sinks of CO_2, efforts should be made to work for the conservation and development of forests, greenery in urban areas and so forth in Japan and also to take steps to conserve and expand forests on a global scale, among others."

3. Methane (CH_4) Emissions

3.1 Projection and Evaluation of Methane Emissions in Fiscal 2000. It is estimated that total methane emissions will be about 1,150 Gg in fiscal 2000, given the expected effect of energy-conservation measures, measures to reduce municipal waste (Table 3) and so forth. This is less than the emissions measured in fiscal 1990 (1,380 Gg); hence the target established in the Action Program To Arrest Global Warming ("The emissions of methane gas should not exceed the present level") is estimated to be achievable.

3.2. Projected Effects of Measures. It is projected that methane emissions will be reduced by about 10 Gg with respect to 1990 levels by mining less coal.

Methane emissions from agriculture are expected to be about 100 Gg greater in fiscal 2000 than in fiscal 1990, judging from forecasts based on the "Long-Term Prospects for the Demand and Production of Agricultural Products" established in accordance with the "Agricultural Basic Law." It should be noted, however, that surveys are currently being

Table 3. Methane emissions in fiscal 2000

Category	Emissions
Energy	1.2×10^2 Gg
Agriculture	8.9×10^2 Gg
Waste	1.4×10^2 Gg
Total	11.5×10^2 Gg

Table 4. Nitrous oxide emissions in fiscal 2000

Category	Emissions
Energy	25 Gg
Industrial Processes	15 Gg
Agriculture	5 Gg
Waste	7 Gg
Total	52 Gg

conducted on farming method that will curb the generation of methane, and experimental research is being pursued with regard to fermentation treatment of animal wastes. Because it is difficult at this time to quantitatively project the effects of these efforts, however, they have not been included here.

As for methane emissions from waste, two municipal-waste scenarios were compared. The first assumes that no waste reduction measures are implemented and that the incineration ratio stays at the current level. The second assumes that municipal-waste is reduced by 30% and that the incineration ratio continues to increase in the line with past trends. Methane emissions would be reduced by some 470 Gg in the second scenario as compared with the first.

4. Nitrous Oxide (N_2O) Emissions

4.1. Projection and Evaluation of Nitrous Oxide Emissions in Fiscal 2000. Total nitrous oxide emissions are estimated to be about 52 Gg in fiscal 2000, taking into account the effect of measures to save energy and reduce municipal waste (Table 4).

This is a slight increase with respect to the 1990 level (48 Gg); further efforts to accelerate development of technology to curb emissions and elucidate the mechanism of nitrous oxide emissions are necessary in order to achieve the Action Program's target, namely: "To the extent possible, the emissions of nitrous oxide and other greenhouse gases should not be increased."

4.2. Projected Effects of Measures. It is estimated that energy-conservation measures included in the Long-Term Energy Supply and Demand Outlook, if they are fully implemented, will effectively reduce nitrous oxide emissions in fiscal 2000 by about 2 Gg compared to the level that would be reached if no measures had been taken.

Nitrous oxide emissions from agriculture are estimated to be roughly at the same level in fiscal 2000 as they were in fiscal 1990, based on the forecasts using figures from the "Long-Term Prospects for the Demand and Production of Agricultural Products." The implementation of other measures to curb nitrous oxide emissions is planned, including promotion of the use of slow-release fertilizer, but they have not been taken into consideration here due to the difficulty of quantitatively projecting at the present time the effect these measures will have.

30

THE NATIONAL ACTION PLAN FOR THE UNITED KINGDOM

Mark Hammond

Embassy of the United Kingdom
Of Great Britain and Northern Ireland
3100 Massachusetts Ave., NW
Washington, DC 20008

The United Kingdom has already submitted to the Conference of Parties to the Framework Convention on Climate Change its national report, detailing how we intend to fulfill each of our commitments under the convention. These include the detailed program of measures to achieve a return of greenhouse gas emissions to 1990 levels by the year 2000, our assistance to developing countries in implementing the Convention, measures to protect and enhance carbon sinks, support for research into climate change, and the promotion of public education and awareness.

These remarks today principally cover the measures we're taking to limit greenhouse gas emissions, but I'll also touch briefly on some other areas.

Our program is based around the precautionary approach inherent in the Convention. The work of the Intergovernmental Panel on Climate Change has shown that the threat of climate change is such that it is appropriate now to take action ahead of unequivocal evidence being established about the nature and possible effects of man-made climate change.

Our program seeks to take advantage of the scope that exists for cost-effective actions now, but it also emphasizes the importance of exploring the full range of policy tools available, including economic instruments, regulatory measures, voluntary action, and public information. Our Climate Change Program was launched by our Prime Minister in January 1994. It aims to return emissions of greenhouse gases to 1990 levels by the year 2000 and sets out details of the measures being taken in the U.K. for each of the main greenhouse gases. The government recognized at the outset that success of the program could not be delivered by government action alone. It depends on the active participation of everyone who makes decisions about the use of energy.

In recognition of that, we undertook a major consultative exercise involving representatives from all sectors of the economy. We published a discussion document early in 1992, and this was followed by a series of workshops and a national conference in 1993. The program was, therefore, drawn up on the basis of the most detailed discussion and debate of the most cost-effective options available and involves a very active partnership between central and local government, the energy and transport sectors, industry, commerce, environmental groups, and consumer organizations.

Evaluating Climate Change Action Plans, edited by White
Plenum Press, New York, 1996

A second key feature of the program is that every sector has a contribution to make. The program recognizes the importance of the cumulative impact of changes in energy decisions across the economy as a whole.

Thirdly, the program is founded on the view that the role of government is to set the right fiscal, regulatory and institutional framework. Our preference is for economic instruments, which work with the dynamic of the market to give appropriate environmental signals resulting in long-term changes by manufacturers and industry, as well as by consumers.

Thus, two key components of our program are the introduction of value-added tax, levied as a percentage of the consumer's bill on all domestic fuel and energy use, and the announcement of a long-term strategy to increase the duty on transport fuels by at least five percent a year above the rate of inflation. This comes on top of the current cost of around $4 per gallon. The implementation of both those measures was confirmed in our annual budget.

The government also recognizes the value of regulation, where it is appropriate and where it is targeted. For example, the program includes revised regulations to improve the energy efficiency standards of new buildings. The program recognizes that the success of the partnership approach relies on effective communication, making sure that everyone is aware of the climate change issue and the scope for taking cost-effective action.

So an important part of the program is the provision of advice and information, mainly through the Department of the Environment's Energy Efficiency Office, and including a major national publicity campaign "Wasting energy costs the earth" with a budget of some $10 million.

The U.K program covers all the main greenhouse gases, but, in common with other countries, the centerpiece is a set of measures for carbon dioxide. The government's energy projections, which were published in 1992, estimated, on the basis of a scenario assuming central economic growth and fairly low energy prices, that CO_2 emissions in the U.K. would increase from 158 million tons of carbon in 1990 to 168 million tons in the year 2000. So the program aims to find savings amounting to 10 million tons of carbon by the year 2000, about six percent of U.K. emissions.

Looking at emissions by end user, industry and commerce account for about 40 percent of CO_2 emissions, transport and the domestic sectors about 25 percent each, and the public sector something over 6 percent. About 1.5 million tons of saved carbon are expected to arise from supply-side measures in electricity generation;through further encouragement of renewable energy sources, combined heat and power. The rest of the savings arise from demand-side measures focusing on improvements in energy efficiency.

About half of these savings will come from the domestic sector, mainly through the implementation of the new domestic energy tax, which we expect to contribute about 1.5 million tons, and the work of a new Energy Saving Trust, which will promote the efficient use of energy and the implementation of energy efficiency measures by householders.

The business sector is expected to contribute about a quarter of the savings, assisted by the Energy Efficiency Office's Best Practice Program and the campaign entitled, "Making A Corporate Commitment," which aims to encourage top management across the board in industry to make specific commitments on Energy Efficiency Programs which bear more than a passing similarity to some of the U.S. programs.

In the transport sector, which is the fastest growing sector of carbon emissions, long-term increases in fuel duties are expected to contribute about 2.5 million tons of carbon saved over the period. A tenth of the savings will come from the public sector, a large saving in proportion to its size. We believe the reason for this is that since the public sector spends other peoples' money, it must aim to make the most efficient uses of resources. The public sector must play a key role in getting across messages about energy efficiency, and it would be hard to do this credibly without setting ambitious targets for government itself. Therefore, the first target government has set itself is a 15-percent reduction in energy consumption from 1990 levels by the end of 1996.

The U.K. program also tackles the other main greenhouse gases, again focusing on a voluntary partnership approach to seek the best opportunities for cost-effective action. For methane, while there are some continuing uncertainties in the measurement of emissions, we are envisaging a ten-percent reduction in emissions from 1990 levels by the year 2000. This will be achieved using measures such as the encouragement of the use of methane for energy generation, new waste management guidelines, and the encouragement of the production of best practice guidelines for several major industrial sectors.

We also expect to achieve a 75-percent reduction in nitrous oxide emissions as a result of the implementation of changes in nylon manufacture and other industrial sectors.

Monitoring progress is a very vital part of our program. Twice a year we publish a climate change news letter, which contains details of the quarterly monitoring of our target. This is an essential part of the process of developing a response within the framework the government has laid out and a very vital part of the partnership approach that we champion in the report.

Some of the measures will have an immediate impact. Others, like the revised building regulations, will have an increasing impact over time. But our monitoring up to now suggests that we are broadly on course to achieve our target for each of the greenhouse gases in the year 2000.

The Framework Convention is based very much on a partnership between developed and developing countries. We are, after all, in this together, and only with a global response can we tackle such a global problem. The Global Environment Facility is already an important tool in assisting developing countries to plan measures to limit their greenhouse gas emissions, where possible, and to protect carbon sinks. We hope it will become the permanent funding mechanism under the Convention.

In our bilateral programs, we are active in many areas relevant to climate change through such areas as energy efficiency projects in India, forestry aid in 18 countries, such as Nigeria, Brazil and Cameroon, and research on the impacts of climate change on coral reefs.

In March 1993, we staged the Global Technology Partnership Conference to help promote the transfer of environmental technologies to developing countries. We are working with a number of developing countries to enable them to complete their own national inventories and plans for climate change.

The U.K. is co-chairman of the Scientific Working Group of IPCC and plays a leading role in the development of climate change science internationally. The government supports this through scientists, economists and energy specialists attending the IPCC's meetings, and we will continue to support the IPCC as it prepares its report for the first Conference of Parties. Internally in the U.K., an Interagency Committee on Global Environmental Change provides the forum for our agencies to coordinate and plan our approach to global environmental research, particularly for climate change. And the Hadley Center for Climate Prediction and Research, which was opened in 1990 by Baroness Thatcher, is a leading contributor to this scientific effort.

Of course, the year 2000 is only a milestone, not an end in itself. The Berlin meeting of the Conference of Parties, to be held in 1995, will consider what more could and should be done in the next phase of the convention. We are committed to continuing to contribute fully to the international process.We are proud that we were the first country to present it's program under the Convention,and that we were the first in the world to do so for each of the main greenhouse gases. We are convinced that only if the developed country parties to the convention give a clear lead and demonstrate what can be done, can we hope to influence others.

Although the U.K.is responsible for only two or three percent of the world's emissions, it's nevertheless vital that we demonstrate our commitment to prompt and effective action and to comprehensive and open reporting under the Convention. One element of this for the U.K. is that we hope to be the first country whose program will be subject to review under the Convention.

Meanwhile, as our program outlines, we are already turning our attention to the longer term options for the U.K. in continuing our precautionary approach to the threat of climate change. Key challenges for us in the longer term will include the projected growth of road and air transport, rising emissions from electricity generation, and how to realize the potential savings in energy efficiency.

In the transport sector particularly, the growth trends pose significant threats to other aspects of the environment, as well as CO_2 emissions. As the recent Royal Commission on Environmental Pollution's report has highlighted, we can take measures that take better account of the environmental costs of transport and contribute to reductions in CO_2 emissions. The debate that is now ongoing in the U.K. will look at options such as even higher fuel taxes, congestion charging or road pricing, improved land planning to reduce the need to travel, technical improvements in vehicle efficiency, and increasing the public's awareness of the need to modify our behavior and lifestyles.

In the electricity generating sector, a major challenge will arise from the fact that some of our nuclear power stations may be approaching the end of their life. We are reviewing our nuclear policy at the moment, and this will make an important contribution to the view we take of CO_2 emissions in the future.

Energy efficiency improvements, which are so central to our program up to the year 2000, will remain an important source of cost-effective savings. Renewable energy currently provides about 2% of UK electricity. Estimates suggest that up to 25 percent of total energy consumption outside transport could be saved by energy efficient investments with a payback of five years or less. The challenge for us here, as it is for many countries, is to discover how best to realize this potential. And clearly, our experience with the current program will be an important source of information of our future strategy.

Our national responses up to the year 2000 mark an important step forward in tackling climate change. Although our national circumstances may be very different, there is a great deal we can learn from each other. And that process, fostered by meetings such as this, is an important part of the global response to the threat of climate change.

31

THE DEVELOPMENT OF AN ASIA PACIFIC REGIONAL RESPONSE TO CLIMATE CHANGE

John C. Topping, Jr.

Climate Institute
324 4th St., NE
Washington, DC 20002-5812

In August, 1994 the results of a remarkable regional collaboration among the Asia Pacific nations were published - in ten volumes totaling about 1900 pages. Beginning in May 1992 a team of over 60 experts from a dozen nations worked under the direction of the Climate Institute to develop studies of climate change implications and national response strategies for eight nations that collectively possess a quarter of the earth's population. Funded by the Asian Development Bank and the Governments of Australia, Japan and Norway, this effort has developed climate change country studies for Bangladesh, India, Indonesia, Malaysia, Pakistan, the Philippines, Sri Lanka and Viet Nam. The country study findings and national strategy recommendations have been endorsed by government sponsored workshops or inter ministerial committees in each of the eight countries over the past year. An Asian Leaders Summit on Climate Change being convened February 18-19, 1995 in Manila by the Government of the Philippines, an Asian Parliamentarians organization and the Climate Institute is seeking to develop cooperative regional action to follow through on these studies.

It is likely that the Asia Pacific region is the area of the world in which policymakers are most concerned about the potential implications of climate change. This unusual awareness is especially grounded, I believe, in a consciousness of the region's historic vulnerability to climatic variability and with it intense rainfalls, severe storms, floods and long dry spells. The climate of much of the region is dominated by the monsoons. Their occurrence is crucial to food production; yet in excess they can cause severe flooding and economic disruption. The El Niño Southern Oscillation (ENSO) phenomenon periodically rescrambles rainfall patterns throughout much of the region thus affecting agricultural productivity and also produces wide fluctuation in fisheries yields. Moreover, the region has been especially vulnerable historically to the ravages of tropical cyclones. It has been estimated that approximately 95% of the deaths in the world from such severe storms over the past two centuries have occurred in the Asia Pacific region, with the bulk of these occurring near the Bay of Bengal or the South China Sea.

For the past several years senior officials of environmental, meteorological, scientific, planning and energy agencies in Asia have assembled on a regular basis to develop

Evaluating Climate Change Action Plans, edited by White
Plenum Press, New York, 1996

regional approaches to the common challenge posed by climate change. These meetings in Nagoya, Japan in January 1991, in Bangkok, Thailand in March 1993 and most recently in March 1994 in Osaka, Japan have sought to identify areas for common action in research, monitoring and technology and information sharing.

The eight country Regional Study on Global Environmental Issues to which I have alluded is an outgrowth of this regional collaboration as are two parallel soon to be published country studies of China and Thailand, both also funded by the Asian Development Bank.

Using climate scenarios developed by the Commonwealth Scientific and Industrial Research Organization (CSIRO) of Australia for the years 2010 and 2070, at low, mid and high ranges, the eight teams of national experts each performed assessments of potential impacts of climate change both in physical terms, e.g. land loss or crop loss, and, where possible, in monetary terms as well i.e. the annual impact on the economy at that future year as compared to a circumstance in which there had been no climate change. Each team also assessed adaptation options, identifying those that would be most cost effective in addressing vulnerabilities identified in the impacts assessment. Meanwhile, all eight participating countries developed inventories of national greenhouse emissions and seven of the eight countries also examined options for mitigating emissions or preserving or expanding greenhouse sinks. One country team, the Malaysian Institute of Economic Research, focused its policy options analysis only on adaptation options.

The results of this Regional Study are quite striking on a number of fronts.

First, even under the mid level scenarios the potential impacts of climate change tend to be quite adverse on balance in each of the eight countries. In developing its scenarios CSIRO relied on the results of four general circulation models – (GCMs) -three high resolution GCMs used by the Intergovernmental Panel on Climate Change (IPCC), UK-MOH, GFDLH and CCC, and CSIRO9, a medium resolution GCM validated in applications to the Asia Pacific region.

The four together necessarily produce a range of results. These scenarios show fairly sharp sub-regional differences and a wide range between low and high scenarios. Temperatures are expected to rise sharply in inland South and Southeast Asia. Should global temperatures rise as now projected, sea level rise is expected to accelerate as a result of a thermal expansion of upper layers of the ocean and a melting of glaciers. The highest projected increases in average annual temperature are in inland South and Southeast Asia. Temperatures during June, July and August are expected to rise only modestly in South Asia as that region is expected to experience more intense rainfall and cloud cover during the wet season.

All four GCMs indicate a sizable increase in precipitation in the South Asia sub-region (from 11 to 59%) during the monsoon season (April to September). This pronounced trend, however, is not evident for Bangladesh. The models show very mixed results during the South Asia sub-region's dry season. Similarly the models show mixed precipitation trends for both the wet and dry season in the overall southwest and Northeast monsoon regions. The projected intensification of monsoons, also supported by the IPCC Scientific Assessment (Houghton, et.al. 1990) could have very large implications, especially for Pakistan and India.

In fact, the Pakistan country study identifies potential intensification of the monsoon and associated flooding risk as perhaps the greatest national vulnerability to climate change. Associated with these changes might also be a greater variability in water flow with the likelihood of drought as well as flood increasing.

In the other seven countries encompassed within this study the most severe impacts were generally anticipated to fall in the coastal sector due to sea level rise related inundation and erosion, saline intrusion into coastal water supplies, potentially increased risk of storm related damage, and climate change and sea level rise driven disruption of fisheries and aquaculture.

The India country study projects that a one meter sea level rise could place as many as 7.1 million people at risk of displacement. A 60 cm sea level rise by 2070 could, the Indonesia study suggests, displace 3.3 million people. The study projects annual losses to the Indonesian economy of over $25 billion due to land loss, sacrificed socioeconomic activity, and loss to the coastal tourism industry. The Malaysia country study projects that accelerated sea level rise will aggravate Malaysia's already severe coastal erosion problem, adversely affecting tourist beaches and coastal agriculture. Coastal mangroves would be seriously threatened by the same sea level rise; yet well before that they are likely to be severely depleted by rapid coastal development pressures. In Bangladesh, even a mid level sea level rise is likely to result in the destruction of the ecologically and economically important coastal mangrove forest, the Sundarbans. A sea level rise of only 45 cm, the mid scenario for 2070, could inundate 11% of Bangladesh's territory, an area now containing 5% of its population.

Sri Lanka, an island nation dependent on tourism and fisheries for a large portion of its export earnings, is likely to find both of these industries quite adversely affected, even under moderate climate change and sea level rise scenarios. The Viet Nam study projects that large scale sea level rise inundation of the Mekong and Red River Deltas could cause large scale population displacement and loss of agricultural lands. A 90 cm sea level rise, the Philippines country study projects, would lead to submergence of small islands, some portions of the reclaimed areas in Metro Manila and Metro Cebu, some districts in the Manila and Manila Bay lake shore area, and the Laguna Bay lake shore areas.

Several country studies highlight a possibly large environmental refugee problem that may ensue from climate change, sea level rise and fisheries disruption. The Indonesia study projects that as many as 2.9 million Indonesians could be displaced by 2010 from sea level rise and that many now dependent on fisheries for their livelihoods could be uprooted by climate induced changes in fishery yields. A major additional impact of sea level rise projected by the Bangladesh study is increased urbanization with many uprooted rural coastal dwellers fleeing to urban areas. The country studies in both Viet Nam and the Philippines indicate potentially large scale dislocations of coastal dwellers from sea level rise. The Pakistan study recommends that high priority be given to the development of a cooperative regional response to the environmental refugee problem.

Although the studies identify many adaptation options such as development of more salt and water resistant cultivars and better water resource management which can certainly be implemented irrespective of climate change concerns, a number of proactive adaptation options are also endorsed. The Indonesia study recommends the construction on an extensive sea dike system to protect such threatened cities as Jakarta, Surabaya, Semarang, Medan and Ujung Pandang from sea level rise. It also recommends the enforcement of a strict setback zone barring development within perhaps 200 meters of the coast in currently undeveloped areas likely to be threatened by sea level rise. The Viet Nam study recommends that threatened Red River and Mekong Delta regions be protected by an upgrading of 2700 km of existing sea estuary dikes and construction of 2000 km of new dikes. The total cost of this construction is projected at US $927 million.

Each of the country teams developed national inventories of greenhouse emissions. In the case of Malaysia the work focused only on emissions from fossil fuel combustion. Table 1 summarized these findings.

Although fossil fuel combustion may account for just over half the combined greenhouse emissions in these eight countries, agriculture appears to be the largest source of emissions in three countries — Bangladesh, Sri Lanka and Viet Nam — and land use change appears to be the largest emissions source in Indonesia.

Table 1. Emissions inventory summary of thematic overview, prepared by Gibbs

Country	Current Emissions (Gg CO_2-Equivalent)	Per Capita Emissions (tons/person/yr)	Comments
Bangladesh	51,389 - 88,048	0.46 - 0.78	Agriculture accounts for about 76% of emissions.
India	809,432	0.93	Fossil fuel combustion accounts for about 79% of emissions.
Indonesia	708,682	3.7	Land use accounts for 72% of emissions .
Malaysia	121,367	7.1	Emissions from fossil fuel combustion only.
Pakistan	114,557 - 128,637	0.95 - 1.1	Fossil fuel combustion accounts for about 55% of emissions.
Philippines	75,196 - 88,638	1.3 - 1.5	Fossil fuel combustion accounts for about 45% of emissions.
Sri Lanka	17,677	1.0	Agriculture accounts for about 38% of emissions.
Viet Nam	84,938 - 112,438	1.3 - 1.7	Agriculture accounts for about 44% of emissions.
Total	**1,941,823-2,033,504**	**1.1 - 1.2**	Fossil fuel combustion accounts for about 52% of emissions.

Source: Country Study Reports.

Four country teams — those in India, Indonesia, Sri Lanka and Viet Nam — developed projections of likely future CO_2 emissions from the energy sector, summarized in Table 2.

There are several reasons for the Asia pacific region's likely substantial increases in emissions. These include rapid economic growth anticipated in much of the region, increasing reliance on coal as a source of electricity, and major increases in demand for transportation services. Aggressive use of conservation measures can, the studies suggest, result in somewhat less steep emissions trajectories. Yet even under these circumstances total energy sector CO_2 emissions are likely to rise steeply through 2070 barring some radical technological breakthroughs. I suspect that these findings in our Asian Regional Study are likely to be mirrored in country study work underway in other populous developing countries of Asia, Africa or Latin America.

For that reason, there may be need for a dramatic reorientation of the current approaches taken in the INC negotiations. Only radical technological innovation that significantly accelerates the arrival of low greenhouse energy, industrial and agricultural technologies can arrest this upward trend driven by population dynamics and development. Schemes based on an elaborate international regulatory mechanism, large-scale transfer payments from industrial to developing countries, or a global carbon tax may be both politically unsustainable and likely to be of little avail in countering this trend.

Table 2. Forecasts of CO_2 emissions from the energy sector (Gg CO_2), thematic overview, prepared by Ebert

Country	1990	2000	2010	2070
India	614,360	NA	1,688,000	NA
Indonesia	120,593	NA	229,625	326,696
Sri Lanka	4,360	NA	18,970	148,060
Viet Nam	19,276	29,935	46,460	NA

NA = Not available

India: 1990 estimate is for 1989-90; 2010 estimate is for 2009/10 and represents Business As Usual scenario. Low conservation scenario is 1,492,000 Gg CO_2; high conservation scenario is 1,355,000 Gg CO_2.

Indonesia: 1990 estimate is for 1988. Significant increases in CO_2 emissions are expected as coal use increases to meet increased energy demands.

Sri Lanka: Value listed as 1990 is for 1991. Forecasts are for the Base Case. Conservation scenario estimated 15,540 Gg CO_2 in 2010; 122,210 Gg CO_2 in 2070.

If China is not ultimately to burn three billion tons annually of coal it will not be because of some international mandate or even Global Environmental Facility (GEF) funding, but rather because some other energy source has become available which is cheaper than coal from the standpoint of the Chinese economy. The same can be said for most other developing countries in Asia or elsewhere.

To meet this need for introduction of cost competitive greenhouse benign technologies some Asian governments have initiated steps to build public-private partnerships to accelerate such innovations. The Climate Institute is now working with a number of energy and financial experts in Asia and the U.S. to flesh out options for consideration by the Asian Parliamentarians at the Asian Leaders summit on Climate Change in Manila.

Such an initiative, if backed by significant resource commitments by OECD countries, would send a clear signal to developing countries of the seriousness of the industrialized world about addressing the climate change challenge. Moreover, unlike major international transfer payments, this initiative is politically sustainable within the key industrialized countries. The same breakthrough in wind energy that might be crucial to a Chinese village could be useful in California; the solar breakthrough that would benefit Mexico, India or Nigeria, might also prove of value in Arizona, Utah or Italy.

In addition to addressing the likely upward spiral of the global greenhouse emission trajectory through the concerted funding of potential breakthroughs in cleaner technologies, serious attention should also be given in the INC process to enhancing the capacity of vulnerable developing nations to adapt to climate change. Many countries, low lying developing island states and poor semi-arid nations, are extremely vulnerable to climate change or sea level rise, yet few of these nations have or are likely to produce a significant share of global greenhouse emissions. It is important, I believe, that the resources of GEF and other multilateral programs as well as bilateral programs give priority to assisting developing countries in fostering credible adaptive responses to climatic variability and climate change. Just as the International Rice Research Institute in the Philippines has played a pivotal role in improving rice production, other world class research and technical institutions could be situated in key developing country regions. Moreover, climate change can often be addressed by more effective responses to climatic variability.

Earlier this year Bangladesh was struck by a severe tropical cyclone with a wind force roughly equivalent to the 1991 storm that claimed about 138,000 lives. Yet, thanks in part to institution of an effective storm warning system with help from the World Meteorological Organization, the 1994 storm claimed only a few hundred fatalities.

Tangible steps such as this, initiated by the first meeting of the conference of the Parties, will engender the confidence of key developing countries that the OECD and other industrialized nations are committed to credible and effective action. This will help also to persuade key developing countries to give increased priority to building energy efficiency and greenhouse concerns into their national development strategies.

32

DISCUSSION WITH RICHARD FINDLAY, MARK HAMMOND, SATOSHI TANAKA, WOLFRAM SCHOETT, AND JOHN TOPPING

Mr. Jhirad: What plans, if any, are being made to address the long-term necessity for significantly reducing greenhouse gas emissions from 1990 levels? To stabilize concentrations of greenhouse gas emissions by the middle of the next century would require significant reductions beyond the goal for 2000. This gets into the whole realm of R&D policy for the long term, but it also gets into the short term. There are many areas: regulations, standards, taxes, physical incentives, voluntary measures and market measures. There are a whole host of these policy instruments, both regulatory and market-based, and some combinations.

Perhaps each panelist could talk about the plans for the long term, address even more significant reductions and give their views on which policy instruments they find are actually effective and are invariant to which political party happens to be controlling the government at the time.

Mr. Findlay: The Ministers of Environment and Energy have asked that the Canadian National Process develop options for reductions beyond stabilization, beyond the year 2000. We have a package of measures which has yet to be modeled in its entirety, but which does provide an opportunity for decision makers to consider measures that go into the future.

Dr. Schoett: From our experience, Germany has always very much trusted in the command-and-control scheme. Since the early 70s we have built up a comprehensive network of laws resting on three pillars: the Federal Emissions Protection Act, the Water Protection Act, and the Waste Management Act, each containing an elaborate set of regulations and ordinances. This legislation has generated a strong demand for environmental technologies, in particular end-of-pipe technologies. By adding these technologies to conventional production processes significant reductions in air and water pollution have been achieved, but at additional cost. Now, stronger emphasis is put on preventative technologies, many of which include intrinsic economic benefits, thus introducing market forces.

The second instrument is research and development of new environmental technologies. The Federal Government has established a comprehensive funding program. The development of new technologies can also serve to improve legislation by setting lower emission limits once a more powerful technology has been demonstrated. We continue to try to introduce voluntary measures by persuading industry to take voluntary obligations to reduce emissions, for instance the automotive or the electronic industry.

Finally, we continue to increase public awareness by information campaigns and teaching programs for professionals and school children.

Mr. Tanaka: Regarding the long-term vision about CO_2 or other greenhouse gases, we have no clear estimation or projection or programs, but we estimate that if the old programs or efforts, listed in the national reports, will be achieved, probably the level of the CO_2 or other greenhouses gases will be stabilized in the late 90s. But it is not a clear estimation, rather desire or hope.

We are considering the possibility of using economic instruments, such as the charges or taxes.

And at the present, we are promoting traditional types of incentives, such as subsidies or a special taxation scheme. And other than that, we are starting and studying.

Mr. Topping: As a practical matter, when you start looking at the numbers, the vast majority of countries, with the possible exception of the U.K., are failing to meet their own targets right now. The numbers of developing countries are exploding, and generally for good reasons — there's a rapid increase in the middle class, and it's really middle-class population growth that drives greenhouse emissions. That's demonstrated in India and China and Indonesia, et cetera.

When we completed, through our local teams, studies in eight countries in Asia; including India, Indonesia and Vietnam, some of them made projections. One projection was, for example, in Sri Lanka from 1990 to 2070, a 34-fold increase in energy sector emissions, and that's even assuming you go heavily nuclear. There's no nuclear there right now.

The implications of this sort of thing are that, barring introduction of radically improved technologies that are cost effective and that are cheaper than the current fossil fuel-based technologies, we're going to explode through the roof. We'll hit the high, high side of the IPCC scenarios, and the effects of global climate are going to be absolutely profound.

With the current discussions, there's a huge amount of controversy over the tiny differences between the environmentalists and the most hard line of industry, maybe 0.3 degree Celsius difference for the next 40 or 50 years. I'm not sure we can afford even those kinds of emission increases given the potential sensitivity and the huge uncertainty in the system.

In a large number of countries industry is represented at the negotiation table by the regulatory managers whose basic job is to show how they block this regulation or they block that happening and so forth.

Somehow we have to transform this discussion so that you bring to the table the engineers and the innovators within industry, because that's the only way it's going to be solved. You have to somehow succeed the way we did in the CFC issue, where we were able to produce very rapid phaseouts, much more rapidly than any of us would have believed possible in 1987, when the Montreal Protocol was negotiated. That was largely because industry was able to work collaboratively and also closely with government. And the rapid phasedowns in 1990 in London and then later in Copenhagen in 1992 were really a function of our ability to bring in cost-effective technological innovation quickly.

Somehow we have to transform this process. If we don't, it's all beside the point, and the discussion in Berlin will just end up with a lot of finger pointing. We very much need to change this and focus very heavily on developing a serious international public-private partnership to bring on cleaner forms of energy a generation faster than what happened in the marketplace.

Mr. Jhirad: Does the U.K. support the adoption of a target for 20-percent reduction below 1990 by 2005? Explain how the value-added tax, the VAT, on fuel and energy and transport fit into the framework of being cost effective.

Mr. Hammond: So far as the question on a 20-percent reduction by 2005 is concerned, I'd refer that to any member of the European Union Commission. Broadly speaking, no, we don't believe that a 20-percent reduction target is feasible between 2000 and 2005, which is not to say at all that if that was the consensus view among Annex 1 countries to the Convention, we would hold out. But I don't believe you will find more than three or four countries who believe that it is feasible for them in that time scale.

On the question on VAT, that applies to domestic fuel, not to transport, which is a slightly separate question. How do we square that with cost effectiveness? If you are looking to reduce carbon emissions, you are inevitably looking at having a serious impact on the amount of energy that is used, either from current levels or certainly from projected levels. Therefore, it is very important to make people think, not once or twice but three or four times, about the energy they use. And the price mechanism is clearly an important way of doing that.

It is, we believe, more cost effective in the end to provide a market signal of that kind than to take money from people through taxes and spend it in other ways, through infrastructure projects or otherwise. This leaves the decisions down to the individual consumers.

Mr. Jhirad: Is the policy of increased living space per capita, as noted in Japan's national action plan, consistent with the global climate concerns?

Mr. Tanaka: Unfortunately, I'm not sure what was discussed about the per capital target and total amount of target in establishing the national program about CO_2. But I have to say now that regarding fulfilling the international commitment to control CO_2 or other gases, both of the two targets are the national commitment of Japan, not only on the per capita basis, but also on the total amount basis.

Our understanding is that the total amount target is very difficult to achieve, but the efforts should be promoted and achieved by the year 2000.

Mr. Jhirad: The other three questions are, "What are specific ways the various countries are trying to communicate to the public the importance of climate actions?" That's question one.

Question two, "What plans do your countries have for meeting your energy needs through nuclear power?" And three, "How much interest is there in joint implementation, and will there be any efforts to have pilot programs like the U.S. IJA, International Joint Implementation Program?"

Mr. Findlay: Joint implementation is very much a part of the Federal Government of Canada's approach internationally. We are encouraging the development of pilot projects, but so far, there hasn't been a lot of action on behalf of Canadian industries to put together projects outside of Canada.

On the nuclear question, parts of Canada do rely heavily on nuclear energy. It comes down to an economic and supply issue as to what the role of nuclear will be in the future. We, in the provinces which have nuclear power, have an over-supply situation now and expect to have that situation for a quite a long period of time. So we don't expect any new growth in the nuclear side of things for the foreseeable future.

Ways to communicate to the public? I think the whole process by which Canada is developing our national action program is a means of reaching out and involving Canadians in this huge issue and recognizing that this is a huge opportunity. This is an issue of sustainability and all these things related to the environment and the economy come together here.

The measures that we are considering include, for example, public education and information outreach programs, and they are increasing well in advance of a National Action Program officially getting under way.

Dr. Schoett: In Germany much emphasis ia put on improving public awareness by training, education and information campaigns. The federal as well as state governments and the municipalities do a lot to spread knowledge about threats to climate and measures for prevention. In schools there are a well established programs to educate younger people and make them aware of the hazards.

Concerning nuclear power, there's been a long lasting debate, in particular after the Chernoble accident, about it's risks. The Green Party is strongly opposing nuclear power and proposes the immediate shutdown of all power plants — without offering a concept of how to reduce fossil fuel burning. The Social Democrats take a more moderate position, but would nevertheless not agree to any new power plant. The acceptance of the public is going downhill. At the moment no new power plant is under construction or even in the authorization process.

Personally, I think that in about ten years, we will have a major change of minds when it becomes obvious that emissions cannot be reduced to an environmentally adequate limit and renewable energies will not be advanced enough to replace fossil fuels. I think we have to be patient.

Mr. Tanaka: Concerning not only CO_2 and the greenhouse gases, but also other pollution problems, such as NO_2 or miscellaneous water pollution, the basic environmental law has a section about the enhancement of public awareness.

And regarding nuclear power, this is about 28 percent of our total supply of electricity. We are assuming that security and safety are secured, and we are promoting nuclear power.

Mr. Jhirad: How about joint implementation with other countries.

Mr. Tanaka: Joint implementation is a very sensitive problem in the International Negotiations Committee, but we are studying with China the possibility of joint implementation. But it has not been clear, like USIJI, with Costa Rica or some other countries, but understand that it's very, very important for the reduction of CO_2 or other gases.

Mr. Hammond: The public information campaign that we've been involved in was, we thought, rather hard hitting, and it broadly consisted of the picture of energy use in the home, a picture of the earth boiling, and a small child's voice pleading with his parents to save the energy, followed by the slogan, "Wasting energy costs the earth." This was considered by our Advertising Standards Council sufficiently hard hitting that they actually hauled us up to explain it a bit more.

But we will press on. On nuclear power, I mentioned we're in the middle of a nuclear policy review, looking at the economics of nuclear power in Britain. I would not say that nuclear power is seen as a policy response to climate change. However, 20 percent of our electricity comes from nuclear power, and if that was displaced in the near or medium term and had to be replaced with fossil fuels, clearly, that's a big issue. The future of nuclear power is being decided on its merits, not necessarily as a response to climate change.

On Joint Implementation, the EU has made its position very clear in the INC and we fully endorse the concept. We recognize the great potential of JI to be a contributor to a global response but we've also expressed some very serious concerns. We believe that the convention will not authorize the use of JI in terms of carbon credits between now and 2000. We believe that there is a very strong case for ensuring that JI only takes place between two parties, both of whom have quantified commitments under the Convention. And we also believe decisions on JI should be taken in conjunction with decisions on future emissions commitments under the Convention.

Mr. Jhirad: I think that, on nuclear power, the answer seems to be that the option should be kept open, for reasons that involve economics, security, public awareness, acceptance and environment locally, as well as climate change.

On education, it's clear that most countries here have a very vigorous public education campaign, and in Canada's case, involve the stakeholders directly.

And on JI, I think the verdict speaks for itself, that it is an option that should be explored for parties who have made commitments under the Convention.

33

CRITIQUE OF SESSION IV

Jennifer Morgan

United States Climate Action Network
1350 New York Ave. NW, Suite 300
Washington, DC 20005

Some of the stories I have to tell are slightly different than the ones that you've heard and my presentation will give you a different viewpoint.

Climate Action Network, just briefly, is a network of over 150 organizations, internationally, that work on environment and development issues that have been key in the negotiating process of the Climate Treaty. We have nodes all over the world. I'm the U.S. Coordinator for the Climate Action Network.

One thing that we've done is an evaluation of the national plans. These are available for those of you who are interested. We have 19 evaluations of OECD countries that were done by NGOs in the countries. So my comments today come from NGOs in the country who are working with their national governments. We will be doing an update of this for INC 11 with some cross-cutting analysis and adding evaluations from central and eastern European countries.

I want to look very briefly at the plans for each country and how they weigh in. I want to bring us back to Bob Watson's presentation and Dan Lashof's presentation as well, as to what these communications are for, where they're supposed to bring us and what the role of the Convention is in all of this.

Some general points here. There's been a lot of discussion about including non-Annex 1 countries in this process. I want to stress the importance of the OECD countries really showing leadership in their national plans, if they want to bring more non-Annex 1 or any non-Annex 1 countries into the process. OECD countries have to prove that their credibility is real in their own backyard, if they want other countries, countries that need plans,and whose emissions are going to be increasing rapidly, to come in and participate. The first step are these national communications.

We fully support the guidelines that have been put forth. We find them very strong, and we commend the countries that have put forth this serious effort to comply with those guidelines. In this panel, the U.K., Japan, and Germany all have put forth very strong plans. Canada is working on their plan now, and we'll be interested to see how it supplements what they had put out before.

But we're talking about two different levels here. I think we're talking about putting together a national communication, which has been done and is very commendable. The inventories are extremely important. But we're also talking about implementing aspects of

Evaluating Climate Change Action Plans, edited by White
Plenum Press, New York, 1996

those communications to meet the goal of the Convention. And that's where I think most OECD countries have fallen short.

I'll present very briefly some specific examples from the national communications of the countries and then wrap up with some political commentary.

You've heard the U.K. target. Their plan is largely focused on CO_2. Two-thirds of the program savings are fairly clear. The third part of the program is a little bit more uncertain. One question I had was on the Energy Savings Trust Program, which is mostly demand-side management programs — where does the funding package come from?

In many of these countries, their measures were already planned and have just been incorporated into their plans. That has been a common critique by NGOs of some plans. In the U.K., the increase in co-generation and the increase in the use of renewables were measures already planned. It is excellent that they're included, but the call of the Convention was to move forward from that.

And in the U.K., the other concern that NGOs have is that beyond the 2000 period, nuclear is coming back in. We've heard that it's going to be looked at on its own merits, but Climate Action Network really doesn't see nuclear as a policy response to climate change. There are many other options that are out there that are much safer.

Transportation in the U.K. I would recommend that the U.K. government look very closely at the Royal Commission on Environmental Pollution's 110 recommendations on its transport study, and implement some of those.

I want to commend the U.K., especially on its education campaign. This is Article 6 in the Convention and it calls on countries to educate – the U.K. has done an especially good job of that.

In Japan, once again, we have a very thorough communication, but one of the problems here is the per capita goal. This does not comply with Article 4-2A and B of the convention and some of the language concerning levels in 1990 is unclear. Japan has the second highest energy-related CO_2 emissions in the OECD, and they have very high fossil fuel dependency and a high aspect of nuclear power.

One NGO in Japan,in looking at nuclear in the future in Japan said that by 2010,the nuclear power supply will increase by 150 percent. That's substantial to say the least.

In Japan, it seems you have a very energy-efficient economy, and, from the assessment of NGOs, MITI doesn't believe that there's potential for future energy efficiency improvements,. They underestimate the potential for demand-side management, especially in the electricity sector.

In Germany, we have the admirable target that has been taken on by them. Part of that will be fulfilled, of course, by reunification and by the shutting down of plants in the former East Germany.

On the coal subsidies issue, I want to point out an interesting aspect because in Saarbrucken, in Saarland, which is in the heart of coal country in Germany, the city has a 20-percent target reduction. Their utility has worked to become extremely forward thinking on solar and other energy sources, even though coal is right there, because they're looking ahead. They're looking for innovative solutions and they are being watched by all of Europe.

The subsidies issue has to be looked at in light of political realities. Here's a situation where you have a high dependence on coal, yet you have commitments and reductions in cities, and Saarbrucken is a specific example.

The critique of German NGOs on the German plan is that there is a very detailed inventory, but there's no outlook of emissions nor projected effects of the measures listed on emissions of CO_2 or other greenhouse gasses.

And in all of these countries, there really is no fundamental energy shift in their energy policy. In order to meet the goals of the Convention now and to move beyond, which is necessary according to Bob Watson, we really need to think about fundamental shifts in energy policy. Of all the OECD plans, I really can't think of one in which the answer to the question in the questionnaire, "Is there a fundamental shift in energy policy," was yes. We're way behind on that.

Some unique things about the German plan that I want to point out- it does support an energy CO_2 tax if it's part of a common EU policy, and it does discuss efficiency standards for cars. I commend Germany for its leadership on the issue of breaking from the EU at the last INC session and putting forth a discussion paper, they are moving forward and thinking about Berlin and what's going to come out of Berlin.

In Canada's plan, as Mr. Findlay says, we have the federal target of 20 percent and the national target, which the provinces support, of stabilizing. I have heard that at the November 8 meeting there was no agreement that a plan, which included voluntary or regulatory and economic instrument measures, should be presented to the COP. Instead, individual Governments will proceed to develop those initiatives they feel that they can live with, and the ministers will meet again in February to determine what the individual programs add up to. It doesn't seem to have as much direction as was hoped for.

So as far as what Canada will take to Berlin, whether that will be a package that includes specific programs to stabilize or reduce emissions with the NGO participants in the process, is in question right now.

The Climate Change Task Force evidently collapsed in October because the environmentalists and industry could not agree. Industry wanted more of a voluntary staged approach. Climate Action Network in Canada wanted a plan to stabilize by 2000 with reductions to 2005. There was very much diversity between the different models.

So there's a great amount of concern on Canada's side and some disappointment, because we heard 20 percent and got quite excited about that potential. The cities in Canada, are really moving forward. Toronto and Montreal have joined in a transportation initiative. So there's a lot of positive action happening in Canada, and we're really interested to see what comes out of this process.

As far as nuclear goes, I believe that Canada is currently negotiating the sale of nuclear plants to countries of Indonesia, China, Thailand, South Korea and the U.S., and that there's groundwork being laid for possible future sales in Argentina and Guatemala. I need not tell you that this concerns us.

Looking at some commonalities to wrap up here, I would say one commonality is that programs that were already under way have been put into a Climate Action Plan. The link between an energy policy and a climate plan doesn't always occur. You all have energy plans, but to have a climate plan, you have to have a commitment to make it part of the Convention.

The transportation sector is vital in all of these countries. Yet the measures, as in the U.S. plan, are weak. It is difficult to implement any kind of fuel efficiency standards, however, the impact of such measures would greatly improve the situation. Again, there are no fundamental shifts in energy policy.

Talking briefly about international activities, all four countries have been excellent in contributing to the GEF. But I would call on them to go a step further and to look at the current language in the Convention that calls for consistency between the Convention objectives and international lending. In other words, the policies of the World Bank, the policies of the regional development banks or the export credit agencies should be in sync with the objectives of the convention.

This is an area that really holds potential for OECD countries to come together through their executive directors and change the focus of the energy policy of the World Bank. Currently, I believe about one percent of the money from the World Bank goes towards renewable projects. They're moving toward it, but they are funding coal-fired plants in India and they're not funding photovoltaic projects.

In order for technology to move forward, you need the market pull. There's a real opportunity for OECD countries not to put conditionalities on non-Annex 1 loans. The idea is for OECD countries to come together and show the opportunity that exists, that these are really potentially excellent, economically viable technologies, and to have a market pull initiative in some of these different PVs.

Recognize, of course, that there are obvious political realities which are limiting. But I really think that we should use these plans that have been put out. We should look at the common measures. We should work together to redirect the World Bank financing, to promote technology, and to demonstrate the commitment. When these countries signed this Convention, they were making a commitment and recognizing the risks that exist meeting the goals of the Convention. Bring in the non-Annex 1 countries, not by worrying them that they're going to have to make further commitments, but rather by assisting them in developing their communications and technology transfer. Involve the states and cities in this process as well.

There are a number of cities internationally that have made a commitment of a 20-percent reduction in their CO_2 emissions. Cities, I would say, financially are as troubled as Federal Governments, and yet, cities like Miami, Toronto, Copenhagen, and Hannover have adopted these 20-percent targets. They need to be supported. They need to be commended, and they need to be included in the process.

I worry about everyone waiting for someone else to jump, everyone waiting for other people, other countries, to make the hard policy decisions that need to be made.

The German paper provides leadership. They took a risk to do that. The last INC session was not easy for the Germans. This is the first U.N. conference on German soil and I need not emphasize that it's very important for them for Berlin to be a success. We all want Berlin to be a success. And as Dan Reifsnyder says, "Well, what's the success in Berlin going to be?" I think that's what we're all trying to figure out.

These national communications are a definite start, but it's time for the OECD countries to stop waiting for each other and move together, recognizing, of course, the difficult decisions ahead.

34

DISCUSSION WITH MARK HAMMOND, JENNIFER MORGAN, WOLFRAM SCHOETT, AND JOHN TOPPING

Mr. Jhirad: I want to give the participants on the panel a chance to respond to the summary of the NGO reactions to the country plans. I think that Ms. Morgan's points were very effectively made about the fact that in none of the OECD countries does one detect any fundamental shift in energy policy, that there are not any relationships developed between energy policy and climate change policy, and that in the international institutions, the financing and lending is still not in sync with some of the areas identified under the Convention.

The question is, "How much of the impacts in 2010 and 2070 are due to population growth and subsidence from economic development?" In other words, looking at the two factors that drive the CO_2, how much is from population and how much is from economic development?

Mr. Topping: In reference to the sea level aspects, there's no question that it would be significant in some areas, certainly in Jakarta, where the subsidence is probably much more a factor in the near term than the global sea level rise.

I think there's no question that population is a factor. One of the things that's very significant in the whole climate discussion is that there's a tendency for the models to be under-predictive because they largely assume the climate is the one variable that's changing. What you find in most parts of the world is that there are a variety of other anthropogenic stresses, and obviously there will also be some technological change in other things that we hope will create some resilience. What we're finding is we're losing soil cover, soil nutrients. We're losing a lot of other things.

When we do the climate modelings for climate change impacts and agriculture, for example, we tend to assume that the only change is that. But if, during the same time period, we're losing significant top soil, et cetera, the whole anthropogenic impact ends up being substantially larger, and the actual stresses in human society are especially large. That's important to bear in mind.

Mr. Hammond: There are a couple of points that were raised on the U.K. report by Ms. Morgan, but given where each of us sit, I found her review a fair one. I think it is quite difficult to distinguish sometimes between policy measures which are adopted specifically in response to the Convention, from the report and policy measures which were being put in place as a response to the issue of climate change in the run-up to the convention and to Rio.

If one looks in the U.K., our original projections back in '89 for 2000 carbon emissions showed a much higher figure. Between then and time the Convention was adopted, our electricity generating industry had seen the way the wind was blowing and had taken significant decisions about the use of coal and the use of gas. Those are valid decisions, and those are contributing to lower carbon emissions, however you classify them.

In terms of the role of the bank and the lending institutions, that is an important factor. Our view is the bank has been improving and needs to continue to improve.

I'm slightly confused. I'm not sure I understand how you cannot have conditionality. If India wants a coal-fired power station and we only agree to fund photovoltaics, I think we are imposing conditionality. That may or may not be necessary, but it's always going to be true for the recipient countries, the bank and the donor countries need to work together on those aspects.

Ms. Morgan: In the energy sector, the one thing that we've called for, and you can agree or disagree with the percentage numbers, was that something like 25 to 30 percent of energy funding in the World Bank go towards these types of technologies. Not that these be the only technologies funded, but that there would be a major shift away from the ones that are currently promoted.

Dr. Schoett: First, I would like to say that this is a brilliant summary, and I would congratulate you. I also felt it was a fair critique on the reports that have been presented.

I certainly would agree that many of the measures that are included in the report have been planned before the convention but this is not a disadvantage. This gives the proof that in some countries already, before the Convention, one was aware of this threat and the necessity to act. Now they are included in the report, but I think we have also put in measures which look to the future.

I also agree that, of course, the industrialized countries have to concentrate and to help the developing countries. The developing countries are the main issue and without solving the problems there, we won't have a solution for our climate problems at all.

Mr. Findlay: Let me join the chorus of congratulating Ms. Morgan. I think it was a very fair assessment of, in this case, where Canada lies. I've been accused often of being the eternal optimist, and I guess maybe you have to try to be that way if you're going to be in this business, and maybe I did oversell the November 8 commitment of ministers. I didn't mean to. I hope that it was honest that this issue is in the hands of decision makers, which is where we expected that it would be at this time.

I think you've portrayed the process as having ground to a halt and collapsed just prior to that meeting, I would characterize it a little differently. We worked tremendously hard over the last year to develop as much consensus as we could amongst all the stakeholders. When you come to the wall, you just can't go any farther. We expected that we would come to that point and we knew that we would have to then hand it to decision makers.

So it wasn't so much a failure as a disappointment that we didn't take it to that next step or steps that I think all of us would have liked to have seen happen. But it's also an endorsement, I think, of the fact that we do intend to try it again, and that's important.

I quite agree with your comment about cities and their importance. And I think in Canada we have not tapped into the positive energy of the movement of cities towards 20 percent and the kinds of measures that are needed. We're beginning to do that. We found in our national process that engaging the cities in this process was actually surprisingly difficult, perhaps because they were actually far ahead of us.

Mr. Jhirad: I want to take a minute to be an advocate and say that in our programs in the U.S. Agency for International Development, we have for some time accelerated our efforts bilaterally and multilaterally to be much more innovative in addressing the energy and environmental problems of the countries we deal with. And being innovative means, for example, scaling up manufacturing production of energy efficient equipment and photovoltaics. We think there's enormous potential for doing that through the GEF and through the IFC. There are ways of connecting private sector transactions with the climate sensitive technology.

For example, we feel that there are many incentives; we're currently working with the bank and the IFC to stimulate joint ventures, to increase production of photovoltaics, for example, from 2 to 50 megawatts a year to 100 megawatts a year; to have project developers who are building power plants in Asia provide a renewable and efficient service element to those projects, because there are 2 billion people who don't have access to any electricity at all.

It's very difficult to tell someone from India or China, who uses one-fortieth of our electricity consumption, that they can't aspire to the same living standards that we do. They want to live the same way that people in the OECD do. But in order to do that, we need an unprecedented technological transformation and innovation which allows them to get those living standards at far less cost to the global environment. That means an extraordinary amount of technological cooperation between the developing and the developed countries, and I think that's clearly the way we need to go, because we cannot deny people sustainable economic growth.

That's the real challenge, to engage our industries and our private sector deeply in that global transformation so we can provide the lifestyles that everyone aspires to, but at much lower environmental cost.

AFTER THE YEAR 2000

Carol Werner

Environmental and Energy Study Institute
122 C St. NW, Suite 700
Washington, DC 20001

Many things are really interesting as we move to this session of the conference: the IPCC and where the international consensus is with the science; looking at the U.S. Climate Action Plan, why the plan is constructed in the manner that it is; and what other countries are doing in terms of the national communications that they have submitted to the International Negotiating Committee prior to the first Conference of Parties in Berlin.

It seems to me that there are some themes that have become quite clear. We have found that even reaching the initial goal of the International Climate Change Convention, for industrial countries to stabilize their greenhouse gas emissions at 1990 levels by the year 2000, is proving quite difficult for many industrial OECD countries to meet.

While we have seen plans that have varied in detail and extensiveness in the kinds of proposals that have been put forward, this is the easy part we've been grappling with thus far, and *it* is proving difficult.

What will be really critical now is to look past 2000 to where Climate Action plans will need to go; what are the approaches, what are some of the tools that will be needed, what are some of the innovative approaches that can be taken, how will negotiations actually take place, will there be new goals set, will there be particular measures that governments will try to take in concert, how will this unfold?

It will be very important for governments to be honest and straightforward with regard to the implementation of their national plans, and what the realistic expectations are for achieving their goals. As the next stage of negotiations unfolds, we will be able to see a credible effort being put forward, and one that says to the entire international community: "We know where we are." We need to address where we are as opposed to goals that may have been set but that are not being achieved right now.

That will be critical in bringing trust into the process as more and more countries become — and need to become — part of the international process and start submitting their own national reports and making their own commitments under the Climate Convention. It will be critical if we expect developing countries to become players over the next years in full participation and partnership with the industrial world.

It's critical for industrial countries to recognize that there are many policies that do indeed affect each other. As we look at Agenda 21 and its call for sustainable development and what that means in terms of the interrelationship of all the sectors of our economies, it

Evaluating Climate Change Action Plans, edited by White
Plenum Press, New York, 1996

is essential that governments and the citizens of those governments understand the interrelationship of different policies in terms of the relationship of energy to climate policy, of pollution prevention to climate policy, et cetera.

Another important area is development assistance, bilateral as well as multilateral assistance through the World Bank, through the IFC, as well as the regional development banks. We have to be very cognizant and careful to ensure that all of our institutional policies are lined up in a consistent way so that they do not send mixed signals or perverse signals. It is very important to remember how integrated and how interrelated our world is. It is the story of the proverbial balloon; that if we take one action over here, we need to understand what the implications of that action are in other places.

Our topic is "After the Year 2000," to look at the process to determine how the U.S. should reduce greenhouse gas emissions in the United States after the year 2000. This is a process called for in the Administration's Climate Action Plan, which was originally published in October 1993. This process is well under way but final recommendations are not ready to be released. They are on track.

Let me tell you a little bit now about the process that is being followed by the White House and by a very extensive inter-agency effort. First of all, President Clinton, as part of the Climate Change Action Plan, called for the establishment of a long-term strategy working group with representation from ongoing administration climate change activities. The process was to develop an integrative decision framework for scientific analysis and for making policy choices on a long-term strategy.

It also was to examine polices that can impact greenhouse gas emissions levels beyond the year 2000, to build on ongoing activities and to develop further initiatives as necessary. It also was to conduct coordinated public outreach concerning the post-2000 emissions reductions effort. The target is to prepare a year-end 1994 report to the President, which will synthesize available results, identify key ongoing work and recommend next steps.

President Clinton directed the White House National Economic Council, the Office of Environmental Policy, the National Security Council, and the Office of Science and Technology Policy to co-chair this long-term strategy working group to examine all budget, technology, R&D, regulatory and economic policies that could impact greenhouse gas emission levels beyond the year 2000. The working group plans to submit a report to the President before the end of 1994.

The group set up is called a Post-2000 Deputies Task Force, which is drawn from the deputies of a whole range of federal agencies. The co-chairs of the force are the National Economic Council, the Office of Environmental Policy, the Office of Science and Technology Policy and the National Security Council, four co-chairs coming out of different entities at the White House.

And, of course, there always has to be a group of worker bees. So there's also been a strategy working group, again chaired by the four entities mentioned before. The members are again agency representatives from a broad variety of ongoing program activities and other initiatives. They have been tasked with getting input on ideas and policy suggestions, reexamining the raft of policies suggested at the initial White House conference on climate change held a year ago, and preparing a draft report for the agency deputies.

Ten subgroups formed to do this analytical work. Originally, the National Security Council was not a part of the work but was added, and it is heading a subgroup that is responsible for looking at long-range international strategy under the Framework Convention.

The other subgroups include one on decision framework, one on buildings,and one on pricing and market incentives. There is a group on non-personal transportation, which is looking at commercial transport, urban and rural mass transit. There is a working group on natural gas. There's also a working group on biomass and renewables, and one on carbon sequestration and forest offsets. The latter looks at soil and biomass sequestration issues overall. There is a separate group on non-energy, non-CO_2 options, and one on global technology diffusion, and, as I mentioned, the long-range international strategy group.

There are some other Administrative initiatives that are related to this kind of work but that are moving forward on parallel tracks. The Post-2000 Group has been looking at what these other initiatives have been doing and later in 1995 will decide how all these efforts should be integrated.

A couple of these initiatives that I want to mention include, first of all, the Administration's Environmental Technology Initiative, which has held more than 20 workshops around the country. The workshops have been an important source of input and recommendations to the long-range, Post-2000 Group.

There's also the President's Council for Sustainable Development. Their report is expected by mid-summer of 1995. And in the Energy Policy Act of 1992, the National Energy Policy Plan, in which Congress directed the Administration to put together an integrated, least-cost energy plan for the country. That work is proceeding at DOE under the coordination of Marc Chupka. It is anticipated that the National Energy Policy Plan will be completed by mid-summer or early fall of 1995.

There's also another process, a parallel process that was called for in the climate plan, which many of us refer to as Car Talk. If you'll recall, when I mentioned the different subgroups, there was one that was formed on transportation, but it was non-personal transportation. That is because there is a whole separate process underway as part of the climate plan process, to look at personal transportation. It is by far the fastest growing source of greenhouse gas emissions. It is being looked at separately because it was not dealt with to any significant degree in the October 1993 Climate Plan submitted by the Administration.

The federal advisory committee established to look at personal transport is focusing in three areas: of vehicle technology improvements; alternative fuels, as well as ways to reduce vehicle miles traveled, whether through land use, other kinds of alternatives such as telecommuting, carpooling or a whole variety of other options.

The charge to the Post-2000 Group, therefore, was to look at all the kinds of actions and policies that could yield large tonnages, for very significant greenhouse gas reductions over the long term. This means looking at significant institutional changes in how we use energy. For example, there are enormous opportunities for buildings to use much less energy than they do now. These kinds of changes need to be institutionalized, and that obviously takes a long time.

These are options at relatively low cost, but that probably would not happen without some sort of a kickstart, some sort of a policy signal or a barrier being removed. And that is the charge of the Post-2000 Group — a big job indeed.

36

SESSION V DISCUSSION WITH CAROL WERNER

Question: You talked about the 20 meetings around the country. What other public input has there been in any of this process?

Ms. Werner: I think that there has been a whole range of informal input through the Environmental Technology Initiative. There also were climate workshops and a variety of consultations. There have been a number of ancillary things that have been going on with regard to the national energy policy process: ETI, and other climate workshops, which have provided significant opportunities for people to submit comments.

Comment: There are segments of industry who will bear a significant role in this, that believe that the Administration is calling for transparency on the part of everything that industry does and is not very transparent in what it does.

Ms. Werner: I am happy to relay that message. It is important to continue to work towards transparency, and I would also encourage you to set up meetings with folks involved, because I think they will be responsive.

INTEGRATED ASSESSMENT

A Tool for Policy Making

M. Granger Morgan and Hadi Dowlatabadi

Department of Engineering and Public Policy
Carnegie-Mellon University
Pittsburgh, Pennsylvania 15213

We've been doing integrated assessment at Carnegie-Mellon for about three years now, so it is appropriate to talk about what we think we've learned so far. I'm going to offer answers to six questions. First, what are the objectives of integrated assessment?

The point of integrated assessment is to put the various pieces of the problem together in order: to keep some sense of perspective since climate isn't the only problem we face; to develop the understanding necessary to support informed decision making; and, to assure that the research we do will support future decision making.

At Carnegie-Mellon, our assessment work has been guided by seven philosophical principles. I'll just tick them off real fast. Uncertainty has got to be the center of focus, because this is a problem with enormous uncertainty. The approach has to be iterative. You're never going to get it right, and as you figure out that some things are more important than you thought and others are less important, you need to iterate and refine. You can't ignore the things that aren't well understood. That is, you've got to avoid the "drunk looking under the lamp for his key" syndrome, because there are important pieces of the problem which we don't understand, and we need to deal with them somehow.

Values need to be explicit, so that other people can use your analysis. Climate needs to be put in the context of the broader sets of changes going on in the world. A good assessment is likely to be a set of coordinated analyses, because you can't use the same analytical tool to deal with everything. Integrated assessment is not a synonym for large scale modeling. And there need to be multiple assessments, because the results from any one piece of work aren't going to be believable until several different groups begin to reach the same sorts of conclusions.

I've summarized the work of some of the groups around the country and the world who are doing integrated assessment. One sees that people's approaches are quite different. There's a very heterogeneous set of activities under way.

Who are the climate decision makers? Well, many of the assessments efforts that I've just been talking about implicitly assume that the world's policy decisions about climate are going to be made just once and by a single decision maker. That is, they are unitary rational

Evaluating Climate Change Action Plans, edited by White
Plenum Press, New York, 1996

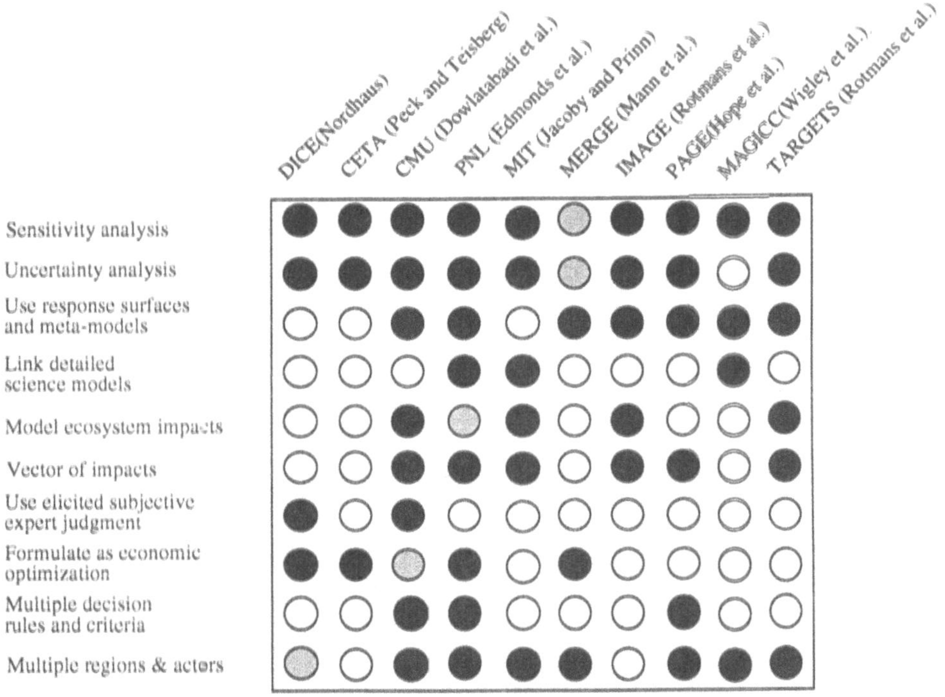

Figure 1A. Comparison of assessments. EPA is developing PEF (Policy Evaluation Framework) but has not released details. Fujii et al. have developed New Earth 21, but I do not have detailed information about it.

actor models. The decision maker might be termed a global commoner who's going to make this rational decision

The people who build these models know that isn't true, but they would argue that such models can at least give a valuable first impression. Our own work has led us to believe that the first impressions gained from a global commoner model may confuse more than they clarify. At least a dozen different nations will make choices which have big climate implications. Many of those choices will not be made by single national decision makers,

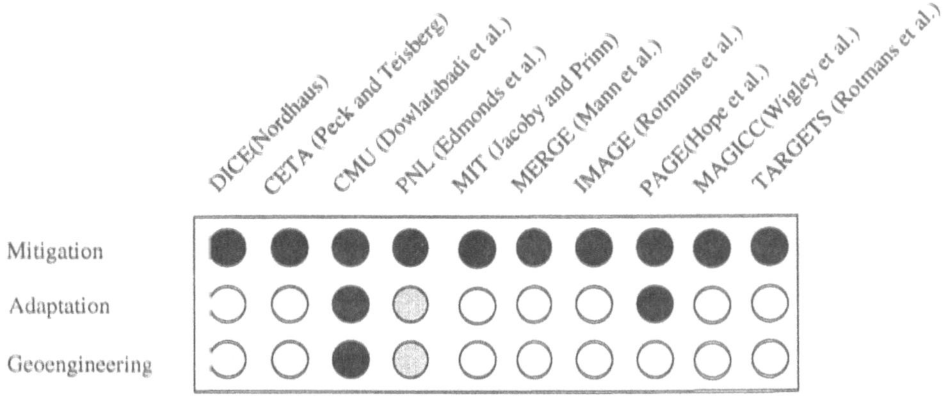

Figure 1B. Policy considered.

but rather through individual choices of millions of organizations and individual citizens and they'll be driven by local interests and conditions. So the first insight on the subject of who's the decision maker is that distributed decision making is one of the fundamental characteristics of the climate problem.

I can illustrate this. We've done a variety of different kinds of analysis at Carnegie-Mellon. I don't have enough time to describe them all, but our current large scale model is something called ICAM-2, Integrated Climate Assessment Model 2. It's a large stochastic simulation. All the variables are treated as uncertain and propagated through stochastic simulation.

With that model, which uses the seven U.N. regions, assume that a single rational decision maker is going to make decisions in each of those seven regions, and might make decisions according to several different decision rules- maximize GNP, go for the largest expected value of GNP, avoid extreme losses in GNP or minimize the sum of GNP losses and ecological losses. What sorts of choices would they make if they were rational actors, given that these regions have different economic situations, different environmental situations and so on?

You run our model, and you get different answers for all these different cells. The point is there are multiple decision makers spread around the world, and they will reach different conclusions.

	Decision Rule			
Regional Decision Maker	Be Risk Taking (Maximize GNP net of abatement and climate impact costs) (discount =5%)	Be Risk Neutral (Maximize Expected GNP) (discount =5%)	Avoid Disasters (Maximize Mimimum GNP) (discount =5%)	Avoid Disasters, Include Nature [a] (Maximize Minimum net GNP) *See Notes Below* (discount =0%)
OECD	Undecided	No controls	No controls	No controls
Former Soviet Block	No controls	No controls	No controls	Stabilize CO_2 Emissions at 1990 levels
China and Communist Asia	Stabilize CO_2 concentrations	No controls	No controls	No controls
Middle East	Stabilize CO_2 concentrations	Stabilize CO_2 Emissions at 1990 levels	Stabilize CO_2 Emissions at 1990 levels	Stabilize CO_2 concentrations
Africa	Stabilize CO_2 Emissions at 1990 levels	Stabilize CO_2 concentrations	No controls	Stabilize CO_2 Emissions at 1990 levels
Latin America	Stabilize CO_2 concentrations	Stabilize CO_2 concentrations	No controls	No controls
South East Asia	No controls	No controls	No controls	Stabilize CO_2 concentrations

Notes:
1) When local non-martket damages and global non-martket damages are equally weighted, and the *expected* value of net GNP is being maximized every region opts for stablization of CO_2 concentrations.
2) At a discount rate of 2% China and S.E. Asia choose stablization of emissions.
3) At a discount rate of 4% the OECD swtich to a no abatement policy.

Figure 2. From ICAM-2.

The second point about decision makers is that these many separate decision makers will each make a long sequence of choices that have climate implications. As time goes by, many of these choices will be designed to adapt to new social and climate circumstances as they emerge. This pattern of sequential and adaptive decision making is an equally fundamental characteristic of the climate problem.

This says that to a greater extent than in any previous environmental problem, the decision makers are a diffuse and often divergent group spread all over the globe, many of whom will make a series of climate-related decisions over a period of many decades. Costs and benefits are both distributed and valued differently among the decision makers, and it's partly for this reason that many of the social and policy tools that we'll need in order to address the climate problem are quite different from those associated with conventional single actor, single decision models that have dominated most public policy thinking to date. In other words, this is a tough problem, and we need a variety of new tools. At Carnegie-Mellon, we're very actively working on trying to devise some of those.

Over the last two days, you've probably heard a fairly complete description of what most scientists and other well-informed people would describe as the climate problem. It's a description that says that human activities result in releases to the atmosphere and they change the radiated balance, that causes the earth to warm up, that causes changes in the dynamics of the ocean atmosphere system, and that causes ecological changes that feed back.

And you have probably also heard discussion about the fact that humans can respond in several different ways. They can abate, that is, they can reduce emissions. They can adapt, that is, they can adjust both their activity patterns and perhaps even their values to those changed conditions. Or they can engage in geo-engineering, which involves modifying the earth's system in various ways.

So that's the climate problem. But is it? In a democracy like the United States, the problem is whatever voters and their elected representatives think it is. So one of the early things we did in our integrated assessment was to observe what lay people know and think about the climate issue.

We've run fairly extensive studies, using techniques developed over the last decade at Carnegie-Mellon. We've conducted open-ended, so-called "mental model" interviews. We learned from those, and from subsequent closed form questionnaire surveys, that respondents regard global warming as bad, and highly likely. Many believe warming has already occurred and 25 percent of our respondents think that in 50 years, there will be over 10 degrees Centigrade of warming. That's real warming!

People tend to confuse stratospheric ozone with the greenhouse effect. They confuse weather with climate. So don't believe things that come from national surveys that are asking questions about climate change, because they may have been read by the respondents as questions about weather change. People do not understand that carbon dioxide from fossil fuel combustion is the main culprit. Thus, they don't tend to identify energy efficiency and conservation as promising mitigation options. People have an erroneously general belief that dirty, yukky stuff causes climate change, and good green practice will make it go away. A variety of specific things, like spray cans, get identified as areas where there are particular problems but spray cans in this country haven't had CFCs for a long time.

When the world is being asked to spend hundreds of billions of dollars on the solution to the problem of climate change, people need a more specific understanding than they have, if they're going to make informed private decisions and if they're going to be informed participants in public debate about the issue of climate change.

The clarifications that are needed aren't that difficult. They're well within the capabilities of modern risk communication methods.

On the basis of what we've learned, we've just produced a new set of public communication materials. It's hierarchically organized. It starts out first with a two-page

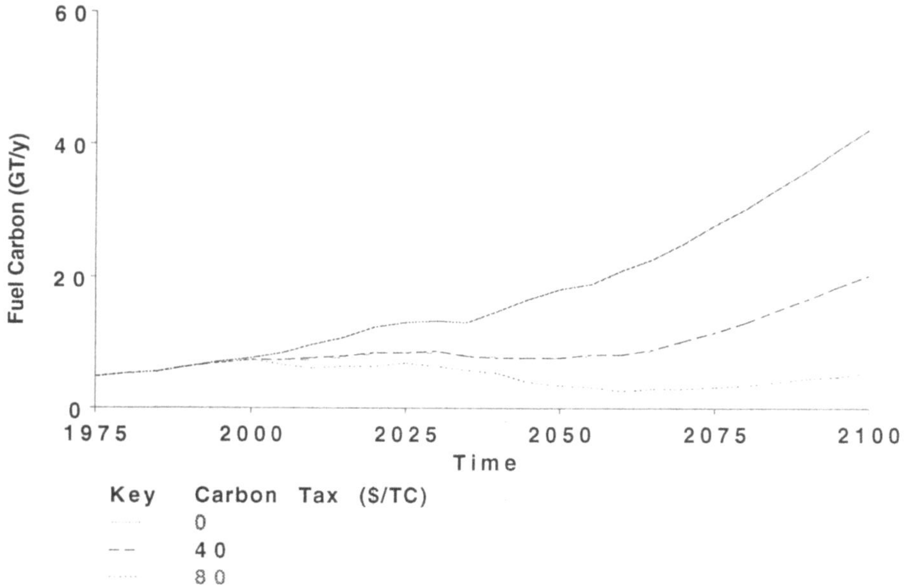

Figure 3. Mean fossil CO_2 emissions.

spread that says, "The issue at a glance. What's climate? What are impact from climate change? What are policy issues?" It also lists and corrects a set of common misconceptions.

Then each of the three questions is elaborated. The climate issue is elaborated with a two-page spread that explores, "What is climate change?" The next two-page spread explores, "If climate change is what might happen." The final two-page spread examines, "What can be done about climate change?" For those who want even more information, there are detailed brochures that pull out of pouches and provide further elaboration. This brochure is available from Carnegie-Mellon.

Turning now to some illustrative results from the ICHM model, Figure 3 reports total mean global fossil CO_2 emissions as a function of time. In this particular case, we have modeled a carbon tax, zero carbon tax, \$40 per ton of carbon tax and \$80 per ton. Carbon emissions continue to rise without tax, and the tax begins to bring them down.

I can then ask: what's the CO_2 concentration that results? It takes roughly \$80 per ton in this model in order to stabilize concentrations.

Incidentally, this is a stochastic simulation. All the variables are treated as uncertain, so I can take that curve and ask: what's the uncertainty associated with it? These are the actual probability distributions here. Given that there's uncertainty about the population growth and there's uncertainty about the economics, and there's uncertainty about the fuel prices, all of these things have been propagated along with appropriate considerations about correlation and the like.

Now, if I'm burning fossil fuels, I'm also emitting sulphur. In figure 5 as time goes by, we see bumps, because oil begins to run out. You have one fuel coming in and replacing another but you can see that under zero carbon, there's a lot of sulphur being emitted. The interesting thing about sulphur is that it's oxidized into sulphate aerosols. Those are very fine, white particulates. They're really very efficient at reflecting solar energy back into space. So if I run the model with and without sulphate aerosols and look at radiated forcing as a function of time, there's a big difference. Aerosols from sulphur have a big effect.

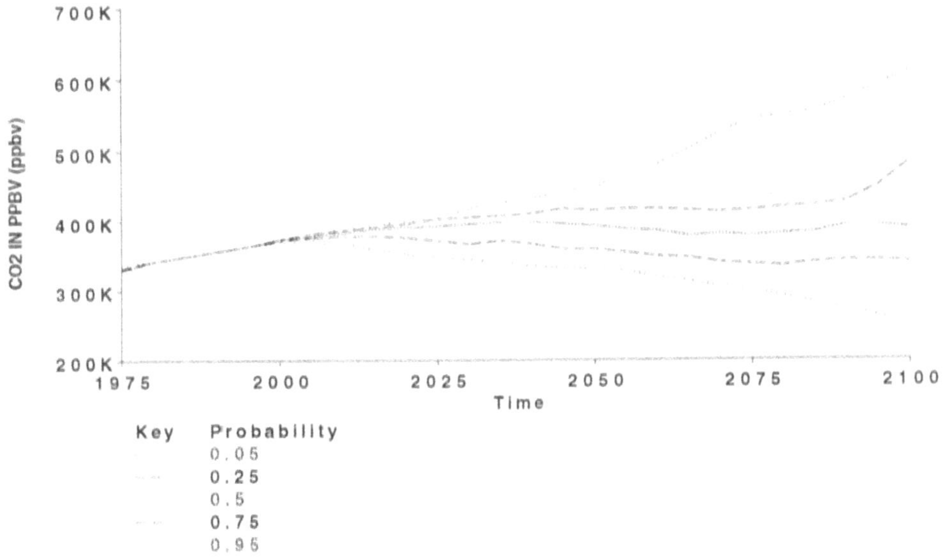

Figure 4. Range of CO_2 concentrations: $80 C tax.

Let's look at the bottom line in terms of the amount of forcing that goes on, with and without aerosols.

In figure 6, if I leave the aerosols out, I get some significant differentiation. At $80 a ton, I stabilize the forcing between 2010 and 2015, figure 7.

Now, suppose I put aerosols in, figure 8. What are they going to do? They're going to delay the point at which the differentiation occurs, because they produce net cooling.

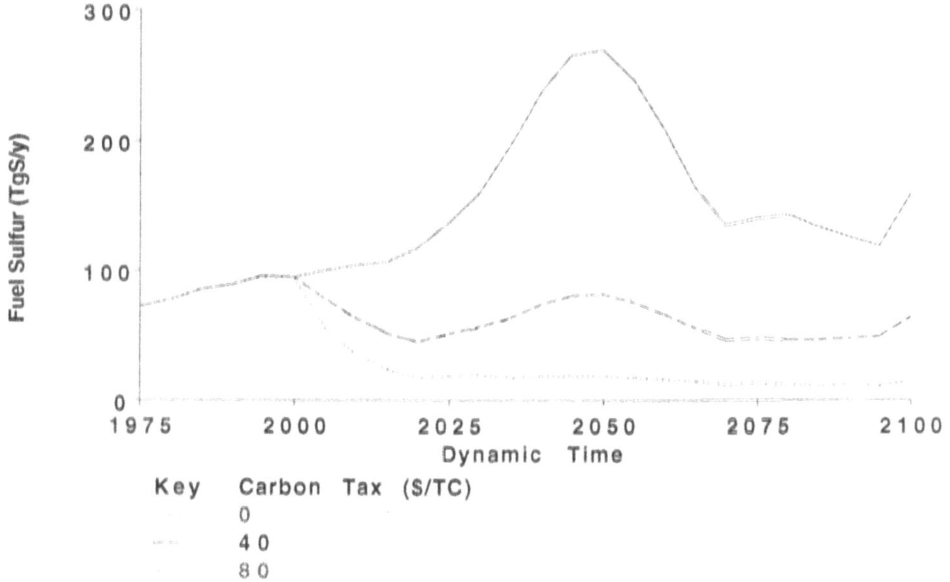

Figure 5. Mean fossil sulfur emissions.

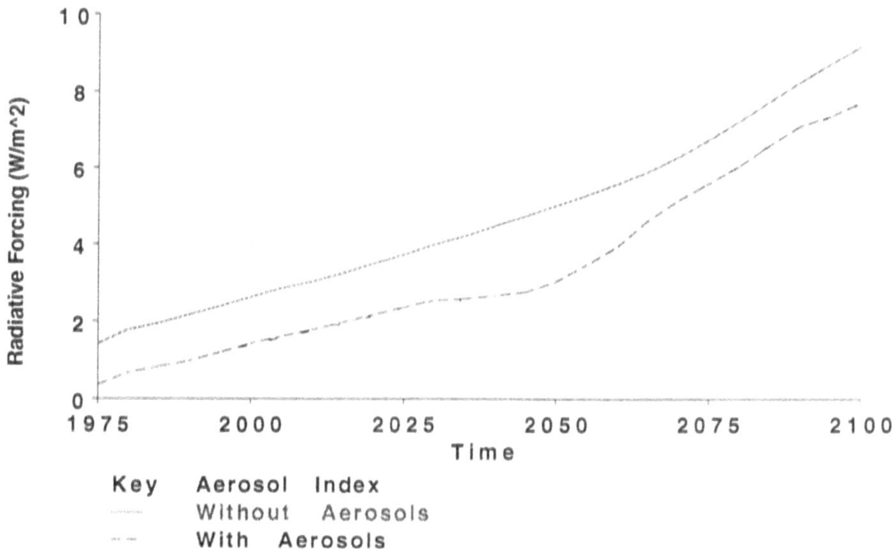

Figure 6. Mean global radiative forcing (GHGS and aerosols).

There's about a 40-year delay before I start seeing a differentiation between these two policies. I suspect people from the White House would tell you that 40 years makes some difference in most public policy debates.

Lots of people have been using a simple index called the global warming potential, to compare methane with nitrous oxide. The GWP says, for example, that per unit mass emitted, methane has about 11 times the impact that carbon dioxide has. You've got to

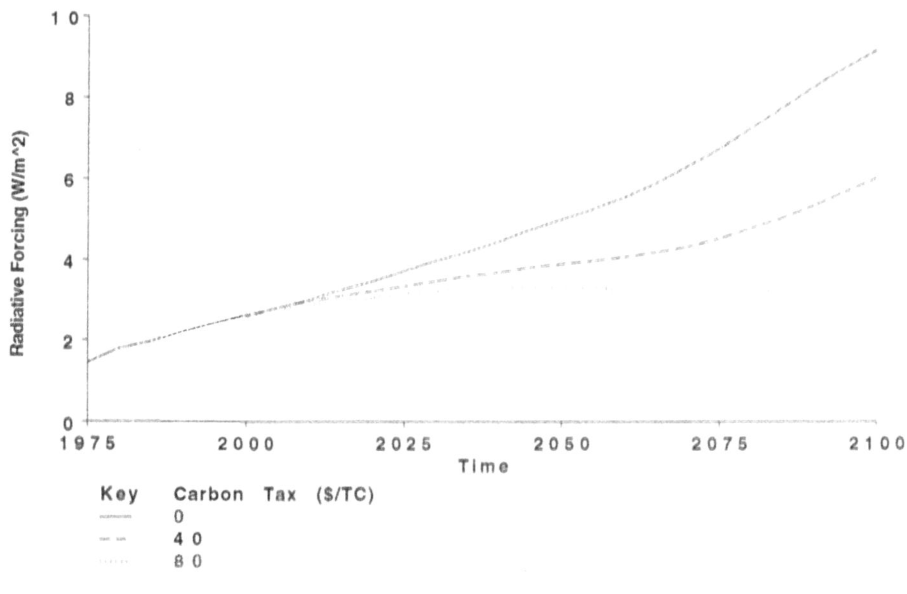

Figure 7. Mean GHG forcing: impact of taxes.

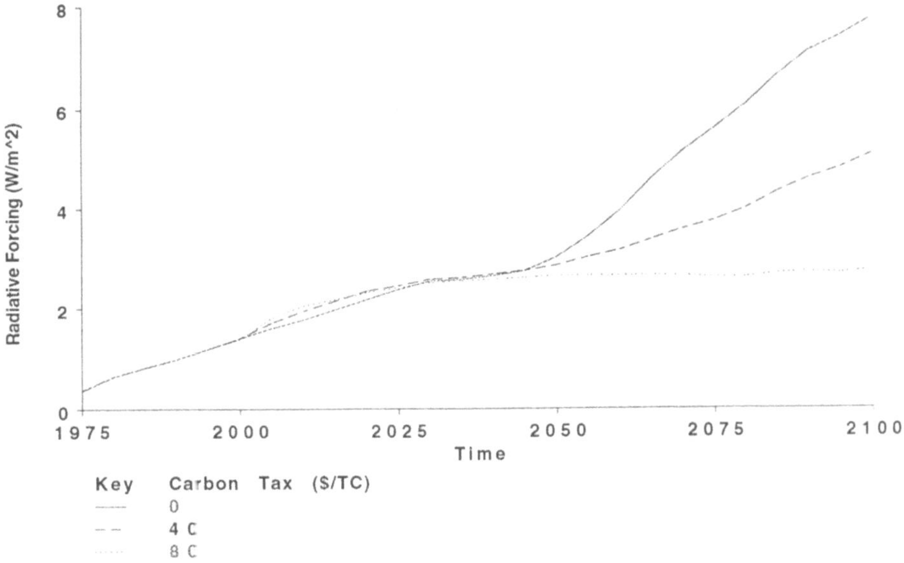

Figure 8. Mean GHG and aerosol forcing: impact of taxes.

compare emissions of methane from developing countries with rice farms to emissions from carbon dioxide from developed countries like the United States.

The problem is that these models are too simple. If you build a scenario-based index, which worries about the fact that different impacts occur at different points in time and that there may or may not be varying degrees of discounting, you can get indices that are dramatically different. There is a factor of three difference in the methane index, if I use IPCC scenario D and a six-percent discount rate. But the main point is GWPs are probably much too simple for negotiating international agreements.

What are the serious impacts? Well, the key insights that I'd offer about impacts are that while the developing world gets hit hardest in absolute and maybe even the percentage terms, the market impacts in the developed world may be larger,but we have much greater adaptive capability. So in terms of impacts on individual quality of life and the fabric of society, it's the developing countries that are most likely to have the biggest problems. The second insight is that people will adapt. I mean, people aren't dumb. If the world starts changing, they change with it. How much adaptation occurs depends on how often they turn over their capital stock, whether they anticipate climate will change in the future when they make their current judgments, and whether they've already detected climate change.

Now, let me illustrate the importance of capital stock turnover with one other illustration out of the ICAM II model. In figure 9 we have total damages versus time. And this is as if capital is very long lived. But if capital is short lived, if I turn over my capital stock and I see that climate is changing, I don't invest in the same stuff. I invest in different stuff. And so it has a dramatic effect on the economic consequences of the change.

We've done a lot of work in eco-system impacts. We've built some very nice statistical models, which divide the globe up into half-degree grids, where we have climate data in each grid. We know the eco-system type in each of those, so we can then estimate a probability surface that tells us, given a particular altitude, a particular climate, what eco-system type to expect. It's clear that different eco systems occur in the same kind of

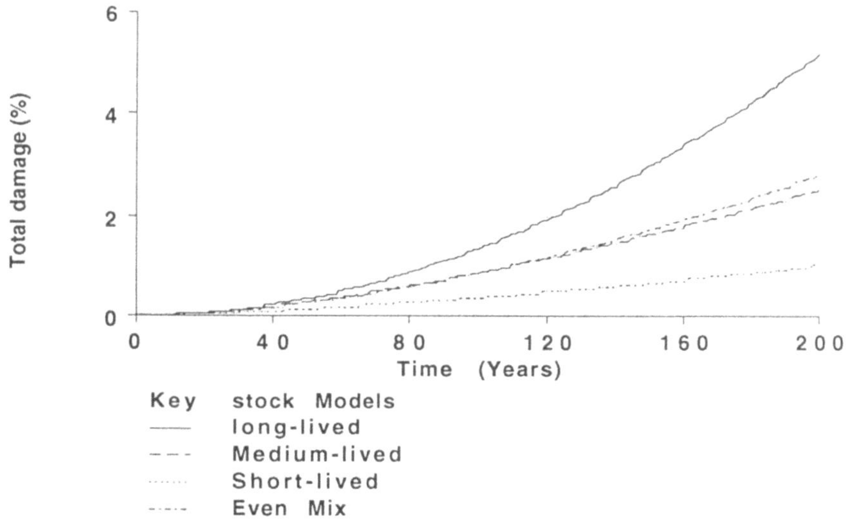

Figure 9. Capital lifetime. The level of damages follow the familiar path if there is no opportunity for adaptation to climate change. This is replicated in the case with long-lived capital. For capital with shorter lifetimes, the damages are significantly lower. The assumed rate of average temperature change is 0.7°C/decade and the damage function follows Nordhaus' familiar figure of .25% of GDP for a temperature change of 3°C.

climatic and altitude zone. And we can use those, then, to do impact analysis and are in the process of doing that.

We are not very happy with the early work on health that's in the literature. We think there's likely to be a great deal more adaptation than, in fact, some of that work suggests.

But, pressing on, what can we say about policy options? I decided to pick just three quick illustrations to give you a sense of the kind of directions we're pushing in. I want to say a word about demand-side management in electricity — that's an energy conservation method — a word about carbon taxes, and then a word about the dread subject of geo-engineering.

The demand-side management literature basically shows two things. The neo-classical economic studies say there's nothing here. If there was something more to be picked up, people would have already picked it up. Engineering studies say that there might be as much as 75-percent reductions possible through demand-side management.

We have looked at all the field evaluation studies of demand-side management that electric power companies and others have done around the country, and conclude that today about one to two percent of peak electric demand is being dealt with through demand-side management.

It's still a little unclear, but results suggest that something like a five-percent saving at a cost of about ten cents a kilowatt hour might be possible, but that if you want a whole lot more than that, it's going to take very dramatic changes of a structural nature and behavioral nature, and it's unlikely that you can gain much, at least on the basis of the behavioral studies completed to date,.

Now let's say a word or two about taxes. Economists argue that fuel price adjustments can be a very effective lever with which to control energy use and hence emissions of CO_2, and that's true. We and some people at RFF have done a series of studies which have looked at this, and the difficulty is that we find some circumstances in which you get very counter-intuitive results. They found a few conditions — certain states in the south — under which the imposition of a carbon tax would actually raise emissions of CO_2. This is due to the high capital costs that are embodied in electricity prices, permitting situations where the relative price rise in electricity generation using coal is smaller than the price rise experienced by consumption of natural gas.

That doesn't mean that carbon taxes are a bad idea. It only says that you have to be very careful about this. You really have to analyze and follow through the implications of all this stuff before you become too enthusiastic. If you put big carbon taxes in place, it has large effects in the long term.

The last policy response that I want to say something about is geo-engineering. It's controversial, and many people don't want to talk about it. In order to stabilize the climate, we're going to have to reach emissions reductions of 40 percent of 1990 emissions and that's a big cut. The prospects of getting there are not all that great. Or even if they are pretty good, and it turns out that climate change is a little more substantial than we thought, there are going to be people around who are going to start arguing that they would like to pursue geo-engineering strategies.

Our group did a fairly large assessment of these strategies, looking at things like wafting foreign particles into the stratosphere. There are some decision makers who are likely to find these sorts of strategies appealing, so our argument is that you really have to have analyzed them ahead of time. You don't want to get to the point where some single country decides to start wafting 12-inch shells into the stratosphere and filling it up with particulates. You need to address the fact that there may well be people who find these approaches appealing, and that to simply forget about them, as opposed to trying to understand them well, is potentially a serious mistake.

What's the state of the climate science? We've recently completed a very extensive set of expert solicitations we've conducted with 16 of the leading climate scientists in the country, very detailed, day-long interviews. Results are now in review at a journal and should be available soon.

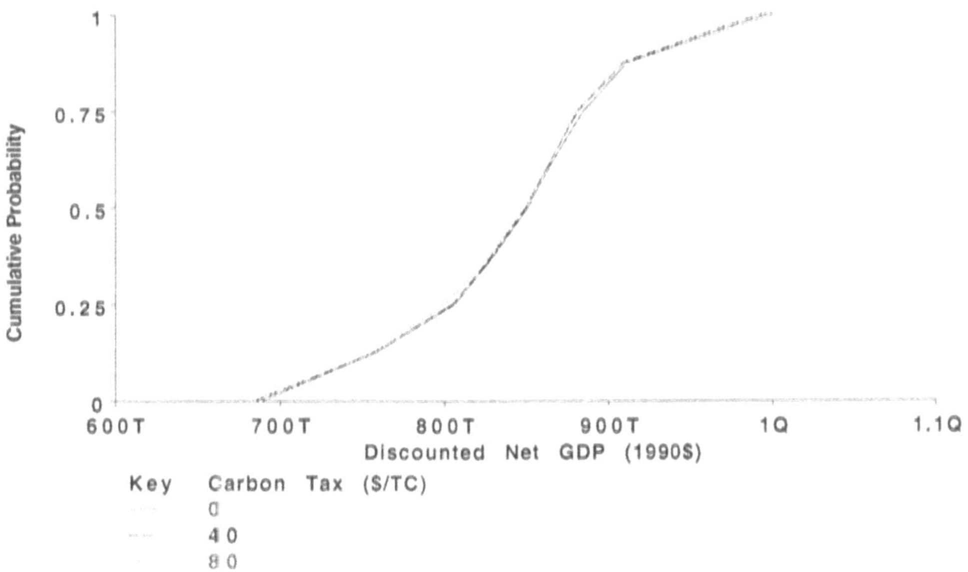

Figure 10. Scientific uncertainty removed: OECD (r=4%).

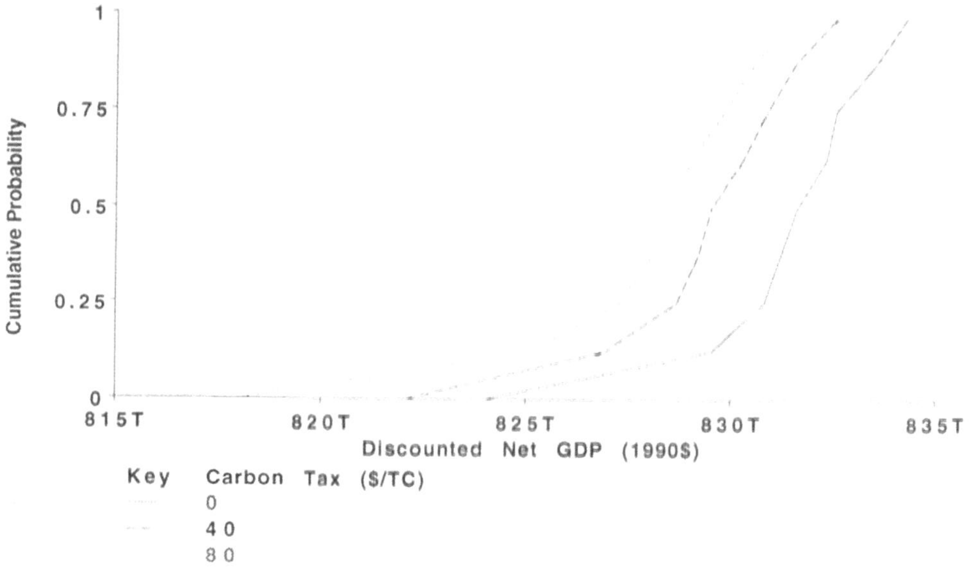

Figure 11. Only impact uncertainty remains: OECD (r=4%).

One of the things we have found in several of our climate assessment activities is that the geo-physics probably isn't the most important source of uncertainty.

I want to show you two final figures, 10 and 11, that show you results from the ICAM II model. These are in the form of cumulative probability distributions for total world GDP. And if you run the model with all the uncertainties in it, and you try to differentiate a policy at, say, zero and $80 per ton carbon, you can't. The two answers fall right on top of each other. Why is that? Because the short-term costs of one are balanced by the discounted long-term costs of the other, and the uncertainties are large.

If I take all the uncertainty about the geo-physics out, and I set all the variables to the mean values, I get the same exact result; that is, I get a big, whopping blobbed distribution. The way to read figure 8 is there's a 25-percent chance that it will be less than or equal 800 trillion. There's a 25-percent change that it will be greater than or equal to 900 trillion and so on. This is cumulative — you integrate the probability density function to get the CDF.

So I can take all the uncertainty out, and I still don't pull the policies apart. I can't tell which one's the right one. On the other hand, take out all the uncertainty about impacts, figure 9 set all those things to their mean values, and things come apart nicely; that is, this policy is clearly stochastically dominant over that one.

That suggests, and many other studies that we've done as well suggest, that uncertainty about the physics is not the only and perhaps not even the most important part of the problem to be working on.

38

INNOVATIVE RESPONSES

Irving Mintzer

University of Maryland
Center For Global Change
Executive Building, Suite 401
7100 Baltimore Ave.
College Park, Maryland 20740

I'm going to leave you shortly to go to Africa to discuss one of the innovative approaches to this issue that I'll present today; that is, the potential for innovative partnerships between commercial enterprises, businesses in the OECD and counterpart organizations in developing countries that will operate through one mechanism of the Climate Convention called Joint Implementation. We're going to try to structure, or at least to identify, ways in which Joint Implementation relationships can promote national economic development priorities in developing countries at the same time that they help us to achieve the overall objectives of the Climate Convention.

But my purpose is largely to do two things. One is to bring you, as a counterpoint to the good news that you've heard so far about the unanimity and consensus on the climate program, a little bit of bad news about how difficult it will be to achieve a solution to this problem in the near term. And potentially how valuable it might be to try, in terms of the opportunities it presents for promoting sustainable economic development here in the U.S. and in the OECD, even if we can't achieve all of our fondest dreams about environmental protection in the next few weeks.

So let me start first with an emphasis on the things where I think we agree about the difficulty that we're facing now. There is a broad general consensus in the scientific community about the physics of the greenhouse effect, particularly the effect of the increase in radiative forcing associated with doubling the CO_2 concentration in the atmosphere.

There's tremendous uncertainty, and that uncertainty will persist, about the distribution of the impacts of climate change at the regional level, in terms of all of the variables that Dr. Morgan laid out. We do not know how much it's going to rain in Washington or how hot it's going to be Omaha or which species are going to do best in central Africa in the next few months or the next few years or, I would argue, perhaps the next few decades — we ought not place all our hopes in resolving these issues and creating substantial unemployment in the community of geo-physicists and climate impact modelers that will result when this problem is solved.

These uncertainties are robust, they're persistent, and they're unlikely to be resolved. What complicates this problem is that we're dealing in effect with the interactions between

Evaluating Climate Change Action Plans, edited by White
Plenum Press, New York, 1996

two large complex, coupled, fluid dynamic systems that are driven by some of the same forces, but have very different response mechanisms; that is, the atmosphere and the ocean. The only thing this tells us in terms of the policy is not that our task simply requires more physicists or bigger computers, but that the surprises and the non-linearities and the discontinuities associated with the interactions between those two fluid dynamic systems are going to keep us guessing about the pattern of the final outcome for quite awhile to come.

So here's the bad news. As Dr. Morgan pointed out, the nature of the climate problem in a democracy is what the President and the Congress and The New York Times and the people in Peoria think it is. In the international situation, the definition of the climate problem is what people can agree it is. In this case, the only basis that we've reached for agreement is a thin consensus called the Framework Convention on Climate Change. That thin consensus has now been signed by about 200 countries and ratified by more than 100. It is an important, but very small, step in addressing this problem. It does not require anybody to reduce anything relative to whatever they did before or hope to do next.

By contrast, the Intergovernmental Panel on Climate Change has indicated that to stabilize the risk of future rapid climate change, that is, to stabilize the concentrations of greenhouse gases, we would have to reduce current emissions of CO_2 and other long-lived gases by 60 to 80 percent relative to their 1990 level. The current Framework Convention has as its fond hope that we will stabilize emissions of greenhouse gases at the 1990 level somewhere around 2000 or maybe later, depending upon how you read the rather convoluted language of Article 4, paragraph 2 and paragraph 3.

The fact that we have not yet agreed on how to stabilize emissions, which will soon lead to a doubling of concentrations, highlights the difficulty that we will have in agreeing on measures, whether they be carbon taxes or tradeable permits or other economically based vehicles for reaching agreement on future reductions.

The one further complication to keep in mind is that not everybody is equally well placed to achieve those reductions. In particular, although the scientists have told us that to stabilize concentrations we need a 60- to 80-percent reduction globally in the emissions of greenhouse gases, the long-life gases, relative to the 1990 level. It's quite clear that the emissions that result from activities that are economically important, that take place in developing countries, are going to increase, regardless of what we say or hope or think or do here or in Geneva or in Nairobi or in New York. These countries are in the midst of an historical transition, shifting their use from renewable fuels, annually cycled biomass, to commercial fuel sources, especially fossil fuels. So their per capita commercial energy use is rising at the same time their population is rising. Not surprisingly, the product is likely to increase when both of these factors go up.

So achieving a 60- to 80-percent reduction at the global level in greenhouse gas emissions would require an even more dramatic increase in the OECD where we do have some scope to increase not just the efficiency of our use, but also the absolute level of our emissions.

We're not going to reduce the absolute level of emissions in developing countries. They're going to grow. The most dramatic thing we might do is slow the rate of growth in emissions, if we are committed to achieving the second equally important half of the objective of the Framework Convention on Climate Change.

The Convention, although it's not often recognized, has two objectives in Article 2. The first is stabilization of greenhouse gas concentrations at a level that will not threaten dangerous anthropogenic interference with the climate system. The second, and in some ways a prerequisite for achieving the first, is that these efforts be done in such a way as to promote sustainable economic development, both in the north and in the south.

Let me just further dampen your enthusiasm in this otherwise hopeful period for all environmentally concerned individuals in our village, that attempting to achieve stabilization of concentrations outside the context of an effort to address the basic human needs of developing countries and the need for improvements in the quality of life and the level of economic development in these countries is simply hopeless; slightly more hopeless than getting all atmospheric physicists and chemists to agree in the next three weeks on the distribution of rainfall or ecosystem impacts as divided by latitude or longitude or any other yardstick you might care to apply.

So that's the context, that this problem can only be solved if closely linked with an integrated, internationally cooperative effort to promote sustainable economic development. That's the rules of the game. That's the bad news.

There is, however, a small shred of good news in this picture. Again, it is emphasis on the fact that all the recent data to the contrary notwithstanding, human beings are smart. We can, under certain stimuli in particularly pressing situations, adapt to change and do things that are constructive, collaborative, cooperative and mutually beneficial.

Those circumstances don't happen as often as we'd like, and there's reason to think that there are structural barriers to intelligent action in many geographic locations. But we can't identify them yet, and we won't go into them here.

What we will say is that the hopeful aspects of this problem involve market-based collaborations, both in the north and in the south, to address the need for profitable economic activity in the short term and environmental protection that has both near-term and long-term benefits.

Here are three quite different examples of the kinds of activity that I think are worth thinking about.

First of all, let's look for just a moment at the situation here in our village and how it is being affected by individual distributed decision making in the broader heartland of America. When I drove down here this morning, the price of gasoline along 13th Street and again along 16th Street was slightly less than the price of bottled water. Now, it's quite possible that there is some basic physical relationship that makes it more difficult to withdraw a liquid that requires no processing out of a hole that's about 200 feet deep than to withdraw liquid that has to be cooked at 1200 degrees, put in a can, hauled almost 8,000 miles, and then put back into a pipe and distributed more broadly. And therefore, there is some basic economic rationality to the pricing strategy that we are now employing. But I wasn't able to see it as I drove down 13th Street.

Nonetheless, we still have individuals who are adaptive and creative here in the heartland of the United States, who recognize that there may be both a short-term profit and a long-term benefit to shifting the balance of our energy system away from carbon-intensive fuels and on to renewable intermittent sources of energy. One of the most promising, and I think one that deserves some special attention, is the work that's been going on at a large, multisectoral energy company, who has recently made a commitment to their stockholders' well-being in the form of a hundred megawatt photovoltaic project that, when linked with their proprietary gas resources, may prove to make both a profitable addition the electricity sector in the United States and a substantial enhancement to their balance sheet that can be viewed in terms of its effect on return on stockholders' equity on their income statement, not in the year 3600, but perhaps before the millennium comes.

The second example in a slightly more stable, mature institutional environment in Costa Rica, where the government finds itself in an awkward and peculiar position, but one which may be of great benefit to those of us here who are concerned about the impacts and the risks of rapid climate change and the need for new promising investment opportunities.

The Costa Rican government, in a period in which the economy was growing and it was clear to all of the economists in the country that everyone was going to be rich pretty quick, made a commitment to protect a substantial fraction, about 20 percent, of the national lands in the form of conservation reserve areas. These were areas that were particularly sensitive and whose assets were particularly valuable. Some of them were to be included in the national park system, others in the national forests. And a third category involved graded lands that had been profitably pillaged for a variety of purposes and were now about to be recovered for the globally valuable purpose of storing what was our waste, carbon dioxide.

When this grand plan was introduced, it was clear that the revenues of the central government were going to increase continuously. Economically good times were here to stay, and there were few uncertainties on the economic side of the picture to be dealt with in policy making.

Not all of those forecasts proved true. In fact, the Costa Rican government has hit a period of hard times, where, surprisingly enough, it didn't have the revenue to buy 20 percent of the country. It had placed easements on this land and restricted its development. But the Costa Rican supreme court threw a startling amount of cold water on this morally quite attractive but economically disastrous scheme in saying that it was quite within the realm of the opportunity and the privilege of the government to place easements and reserves on land. But if it were going to do it in a democracy that was market based, it was startlingly enough going to have to pay for the land and compensate its current owners. And if it didn't have the money, it was going to have to remove the easements and let people carry out whatever process of development they intended.

That nasty, unanticipated outcome has changed the strategy that the Costa Rican government is seeking to employ to serve all of our global needs. And indeed, what they are now proposing to do is an unusual form of joint implementation, in which the government of Costa Rica would establish a portfolio of land management programs whose purpose would be both to protect bio-diversity, as well as to sequester carbon, a form of limited geo-engineering, I would suggest.

What the Costa Rican government would commit to do was to manage these lands on a sustainable basis, to periodically evaluate their carbon storage capacity, and then to sell to interested investors the prestige and the credit for this carbon storage scheme by buying shares in this portfolio that would be called "tradeable deposition rights." These leases on the carbon uptake capability of these lands could be freely traded, acquired by companies in, for example, the U.S. or Japan that needed a domestic credit toward their tax problems and liabilities, would not require any internationally accredited JI scheme.

By holding the lands in a portfolio and only selling rights to a fraction of them, the Costa Ricans would be able to ensure against the unanticipated failures, however, improbable, that would be associated with an occasional forest fire or a bad bug that ate all the crops in a particular area or a small virus that undermined the grass that was taking over a formerly degraded land area.

So by only selling the rights to, say, 50 percent of the carbon storage capability, or perhaps 60 percent of the carbon storage capability, the government would ensure against these nasty surprises and allow the investors to be reasonably certain that their hoped for carbon sequestration rewards would be achieved.

In addition, by managing the lands at the federal level in Costa Rica, they would insulate the investors from any of the odious responsibilities of direct management, including the potential for political embarrassment that might result from some of the other schemes for carbon sequestration, such as carbon cemeteries that have been proposed in Guatemala and other places, where a U.S. company, for example, might just buy, say, a quarter of Guatemala, remove the nuisance people, scrape off the top, plant a set of trees that we could let grow for a couple of centuries, and keep the carbon credits, maybe even offering an employment program for the under-utilized and under-employed members of the former Guatemalan state militia, who could now be hired to walk the fence line and keep those nuisance people from messing our trees while they soak up carbon and protect from global climate change.

The Costa Rican scheme offers us a level of political risk insurance that is attractive for a variety of reasons even beyond those associated with not having to deal with those nuisance people and keep moving them out of the cemetery. It also allows for the transfer of know-how in sustainable management without requiring U.S.-based companies to take a large position on Costa Rican lands that they may not want to end up holding 5, 15, or 50 years down the road.

The third type of scheme that I think is worth thinking about is Joint Implementation that doesn't involve either nurturing trees or converting large capital equipment and carting it to some far-away place where you could drop it by parachute and hope it would do a better job than whatever they were using there before. This third scheme involves the transfer of something slightly more lightweight than electric generators and other power system equipment. It involves an effort by a number of Japanese companies to invest in their Asian colleagues in the form of a transfer of information and management systems for electric utilities.

You know, for example, here in the U.S. and in Japan, it takes about two kilowatts of installed capacity to deliver a kilowatt of average electric generating capacity. In a number of developing countries, in India, in China, in Burundi, it takes four, five, six or seven kilowatts of installed capacity to deliver a kilowatt of average capacity, partly because of the know-how and management expertise in things like the transmission of distribution system and the desire to perform regular maintenance, like oiling the turbo generator.

So what these Japanese companies are beginning to do is to provide tiger teams of trained management specialists, who can go into electric utilities in less-developed countries and help them to identify management and operational problems, whose rectification can result in substantial increases in efficiency without requiring the air drop of thousands of tons of new equipment.

By transferring information, management know-how, and sophisticated techniques of operating systems that are already in place, we may be able to achieve, by cooperative action, the kinds of near-term reductions in emissions that would be hard to achieve and hard to finance if what we had to do was tear out the equipment in place and bring a better mousetrap, of our design, of course, for their local integration and use.

Well, these three examples, then, are where I'll leave you with hope. By investing collaboratively, trusting our partners to manage their own lives, transferring information instead of capital equipment, and doing things here at home to affect the kind of long-term transition that will allow us to make money by doing what's smart, as well as what's right, we may be able to form the kinds of innovative partnerships that can promote sustainable economic development here in our village and across our little heartland, as well as to set an example of how, through enhanced trade, equitable relationships, and collaboration internationally, we can have the opportunity to maintain a fond, if diluted, hope that the objective of Article 2 in the Climate Convention might not prove to be all fantasy.

39

DISCUSSION WITH IRVING MINTZER

Ms. Werner: On the issue of your upcoming trip to Africa what do you think of the potential to reduce greenhouse gas emissions in Africa, with the exception of South Africa and Egypt? Per capita, CO_2 emissions in Africa are the lowest in the world.

Dr. Mintzer: Taking the continent as a whole, the prospect for reducing CO_2 emissions in Africa is almost zero except in absolute terms, assuming, of course, that we will be able to contain the AIDS virus, we'll be able to feed everybody adequately, and we'll be able to maintain peaceful relationships among the inhabitants. As long as we can do those three things, the technical and engineering problems of reducing CO_2 emissions are basically insoluble.

We're not likely to achieve all of those three goals in a week, and the negative impact may be that due to reasons that have nothing to do with concern about climate change, we may allow the situation to deteriorate seriously in a way that undermines not only our hope for peace in Africa, but our expectations of enhanced development and trade with African countries.

There are a couple of examples within the continent, if you look more specifically, where there are real opportunities at the national level to reduce emissions that have not yet been fully tapped. One of the most important is Nigeria, where petroleum prices and gas prices remain quite low, attempts to improve efficiency remain very sparse, and the degree of social coherence and institutional collaboration is still rather limited.

If we can help the Nigerians to see their way clear to promoting economic development through more rational energy pricing, perhaps by doing something as radical as introducing full-cost energy pricing in the U.S., we should be able to do quite well at reducing the emissions of CO_2 in Nigeria.

It will take, however, leadership in one other ancillary field, and that is finding ways to meet the needs for personal transportation other than sending all our old cars to Africa. The Cubans have a limited absorption capacity, and we're going to have to try some other strategy if we don't want emissions in the transportation sector to go through the ceiling.

But more broadly, in Africa there is another important and strong possibility that might be worth addressing. Unfortunately, it involves the unpopular approach here of supporting such radical things as U.S. bilateral assistance and the more constructive use of multilateral development banks. But we could actually try to promote economic development in Africa in ways which maintain soil fertility, minimized erosion, provided adequate food for people, and in so doing promote carbon storage in soils, as well as the efficient industrial development of urban societies in Africa, so that people could have the same basic conveniences that we've become so accustomed to.

There are real possibilities. They involve, unfortunately, a level of political courage which might be hard to find here in the next month. But we could do it if we wanted to.

Ms. Werner: How significant a contribution do you believe that joint implementation projects can make to reducing CO_2 emissions overall?

Dr. Mintzer: Small in the foreseeable term. In fact, if we try to do anything other than achieve a relatively modest level of reductions through JI, we will almost certainly guarantee the entire regime will fail. I would suggest, for example, that we might want to think about an informal international agreement in which OECD countries, at least, might agree to attempt to meet no more than one to three percent of their national emissions reductions targets through emissions reductions that take place in somebody else's backyard.

Still, one to three percent of 100 or 120 million tons in the U.S. is more than enough to buy lunch in a lot of places. It's more than enough to advance the development of some very promising technologies and to establish market niches which could be allowed to grow without the international subsidy associated with joint implementation.

So I would suggest that we aim for something like one-to three-percent limitation on emissions reductions through JI in the first three to five years of any international pilot phase, maybe allowing it to grow, if it's successful, rapidly reaching by the year 2000 something that might approach, say, five percent or seven percent of the total. And after the millennium comes and we have our other problems solved, we could think about setting a goal for joint implementation that might be as much as, say, ten percent of the national emissions reduction target. That still represents several hundred billion dollars worldwide, more than enough to fertilize the market for a variety of attractive technologies.

Ms. Werner: What impact do you believe that GATT will have on global climate change, particularly since you had mentioned world trade?

Dr. Mintzer: GATT has a potentially very positive effect on global climate change, but it, unlike the physics, is highly uncertain. The potentially positive effect of GATT is that it could be, if we chose to make it so, a vehicle for more free open and equitable trade relations between the signatories, potentially between north and south.

If we choose to use trade instead of aid as the vehicle for establishing the momentum for sustainable development, there are very large potential benefits, not only on the environmental side but on a number of other parallel fronts whose successful progress will enable the kinds of peaceful relationships that make environmental progress possible.

If we don't exploit GATT in a way which rigidifies existing trade balances, we have the potential to do these things. Unfortunately, one of the weakest and least defined parts of GATT is the way that it will specifically address environmental issues. And indeed, the WTO has pretty much so far treated environmental issues as kind of a stepchild that we'll get to after we solve the world's economic problems.

I hope I'm there when they call after they solve the world's economic problems, but I might not be young enough to still care a lot about it at the time.

40

SETTING GOALS UNDER THE U.N. FRAMEWORK CONVENTION ON CLIMATE CHANGE

Jonathan C. Pershing

Office of Global Change
Department of State
2201 C St. NW, Room 4333
Washington, DC 20520

In this talk, I would like to do two things: (1) review country activities with respect to their reporting requirements under the FCCC, and (2) provide an overview of the United Nations Framework Convention on Climate Change (FCCC), and some (perhaps notional) expectations for its future functioning.

NATIONAL COMMUNICATIONS

Let me start off with some brief comments regarding the preparation of national communications. In my comments, I hope to offer not so much a detailed view of how countries' communications have been prepared as an overview of the process -a more holistic approach to evaluating all the communications.

The FCCC calls upon "Annex I Parties"[1] to submit their communications within six months of entry into force of the Convention for that Party. Annex I Parties are industrialized country Parties, and Parties from the former Soviet Union and Eastern Europe. As of January 1, 1995, 15 Parties[2] had submitted reports, although 18 communications were due.[3] There are a series of additional national communications that will come due over the next several months.

The United States has begun a preliminary review to look at the substance and form of the various national communications. Our review indicates that in the plans that have been submitted so far, there are more than 700 individual measures. While these are not unique measures - that is, countries often undertake similar programs - they do add up to a remarkably large total.

Almost all national plans include measures addressing the major greenhouse gases: carbon dioxide (CO_2), methane (CH_4) and nitrous oxide (N_2O). Some also include measures to address chlorofluorocarbons (CFCs), while others include other less commonly emitted greenhouse gases, for example, the perfluorocarbons (PFCs). Most plans also include

Evaluating Climate Change Action Plans, edited by White
Plenum Press, New York, 1996

national programs to address sinks of greenhouse gases. Few of the reports include discussions of countries' actions with regard to adaptation. Only approximately one third of the plans provide detailed information projecting the total effects of measures on greenhouse gas emissions, and only the U.S. plan tries to provide some estimate of the impact of each individual measure.

Notwithstanding what information countries chose to provide in their communications; the international community, in the course of negotiations, had reached substantial agreement with respect to what countries *should have* reported. At the 10th session of the Intergovernmental Negotiating Committee, countries adopted reporting guidelines. It is unfortunate that so few countries have fully complied with these guidelines.

COMPLIANCE VS. NON-COMPLIANCE

The lack of compliance raises an interesting – although potentially difficult – problem for the negotiators: What action should be taken when only half the countries that are required to submit plans have done so, and few if any that have submitted plans have met the required guidelines? Dan Reifsnyder[4] in his presentation on the first day of this session raised a couple of the options which might be chosen to look at noncompliance issues; I would like to offer my own perspective here. In my mind, there are two alternatives: either a strict regime, or a more flexible one. The Parties (to date) have opted for the second alternative, foregoing the option of a very strict regime with penalties for noncompliance.

I believe that with the current groundbreaking reporting effort that the FCCC has inaugurated, this second more lenient approach is warranted. In such a mode, Parties will collectively assess, on the basis of the agreed guidance, the adequacy of countries' communications. With the technical assistance of the FCCC Secretariat, this assessment would examine how much work Parties have undertaken, how legitimate the effort has been, and how closely Parties conform to agreed guidelines. In addition, the assessment would consider what countries might do in the future to improve their communications. Based on this knowledge, Parties can then – both individually and collectively – further elaborate on guidelines and prepare future more comprehensive and appropriate national communications. There would be no sanctions for falling short of still ambiguous expectations at this time.

By way of explaining why I favor this more lenient approach, let me first characterize what I believe countries are doing by preparing their reports – and what I believe the negotiators, in adopting the FCCC believed would result from this approach. The underlying goal of developing a report is to assure that each country is fully assessing the actions it is – and could be – taking to mitigate and adapt to climate change. Based on this understanding, the purpose of the review is then two-fold: (1) to assure that minimum standards are followed, and that no country is ignoring their obligation to take actions and report on them, and (2) to provide a comparison of actions to help all Parties evaluate which ones could be most effective at reducing emissions.

While it would be possible to set up a system in which "inadequate" reports would be subject to sanctions, I suggest that this concept would run counter to the intent of the Convention as it was finally adopted. After all, there are no hard and fast commitments in the climate change convention that require countries to meet certain reductions in their greenhouse gas emissions by any date certain. Absent such a signal, against what would reports be measured? We are only now developing a regime for reporting, and to set absolute standards — and sanctions for non-compliance with those standards — until a clearer sense has emerged of what constitutes an acceptable report is at best premature.

However, in the absence of non-compliance procedures, the international community runs the risk that reports will be too brief, and not provide either the incentive for or an accounting of countries' efforts to reduce greenhouse gas emissions. Currently, all a country must do for its report is to prepare an inventory of greenhouse gas emissions and to make some listing of the various measures and programs and policies being taken to try and address those emissions.

Having noted my own preferences, let me remind you of the state of play: The guidelines that were adopted at the INC are not binding. Any measures or resolutions or decision taken by the negotiators prior to the meeting of the First conference of the Parties in Berlin (scheduled for March 28 to April 7, 1995) cannot be binding – they are only recommendations of the preparatory committee to the Parties. So, until the guidelines themselves are adopted (and I think they will be), it is impossible to penalize any country for not having met them. Even if we are to adopt these guidelines, it seems unlikely that countries will approve a more stringent regime in the longer term; countries do not seem willing to submit themselves to the kind of binding and potentially intrusive and controlling review of their communications that may be part of a much more stringent process.

RATIONALE FOR REPORTING

One of the reasons for preparing communications is to provide an opportunity for countries internally to evaluate and review actions being taken to address the climate change problem. Through a review and careful analysis, in some cases a country may discover actions of which it is unaware - and which for climate change or other reasons, it may choose to remedy. Remedies may be completely unrelated to the climate convention's goal, but still may have very substantial implications for modifications of country greenhouse gas emissions. Very few countries ever undertake periodic reviews of domestic actions with a particular aim like this in mind; the FCCC causes such reviews to happen.

As noted above, another objective of the preparation and international review of communications is broad sharing within and across countries of actions that are underway. In other words, it's an information exchange opportunity. In order for such an exchange to work, some synthesis and evaluation of the communications is necessary. Furthermore, it is necessary to convey this information — in a readable manner — to policy-makers. A 200- or a 300-page document or multiple documents will not be reviewed, and is unlikely in an unabridged form to be useful.

The international community is now engaged in this exercise. Once the guidelines for the preparation of communications were developed and adopted, the INC turned its attention to the review process. At its tenth session, the INC adopted a set of recommendations, with components much like those I outlined above. The INC review is being coordinated by the INC interim secretariat, and is expect to produce a synthesis document for the eleventh session of the INC (scheduled in February in New York City). As currently envisioned, this synthesis is not to be a detailed review of each communication, but rather an overview of all of the communications. It is envisioned that following this broad perspective synthesis, there will be an in depth examination of each national communication.

The United States has volunteered to be among the first countries to be reviewed. In the U.S. view, transparency in the reporting and evaluation of communications is critical; we would like to see a similar degree of transparency in the preparation and review of all countries' communications – a facet which, unfortunately seems all to often to be lacking.

The final purpose for preparing national communications and for causing them to be reviewed is to subject countries — both internationally and domestically — to public scrutiny. While this may be a more influential component of the reporting process in the

United States than it is in many other countries, it still serves as a driver to force action. Countries are seldom willing to report in a public manner that they are undertaking few, if any, actions to address a problem. The requirement that countries make public their communications, and open their review process to international scrutiny, inherently effects the policies themselves.

As a forcing agent, this transparent, participatory and comprehensive reporting process ultimately may prove to be the pivotal contribution of the FCCC. The convention's lack of targets and timetables and its hortatory aim were extensively debated. Ultimately, the negotiators decided that the only binding requirement would be for countries to report on their actions. In the near term (i.e., on the timeline of the FCCC negotiations) this may have been the maximum agreement achievable. In the longer term, however, the process of establishing reporting requirements can be expected to be critical to the successful resolution of the climate change problem. Because of the importance of national reports, it will be necessary for the international community to be vigilant about regularly updating and revising the reporting process.

NON-ANNEX I COMMUNICATIONS

Most of the discussion above has focused on the preparation by Annex 1 parties of their national communications. However, all parties have reporting requirements. The convention text addresses reporting in two Articles: Article 4, which talks about commitments for reporting, and Article 12, which talks about the methodologies for preparing reports.

Article 4 has several paragraphs of note. One is paragraph 4.2 in which the requirements for Annex 1 parties to prepare communications are discussed. The second is Article 4.1, which is applicable to all convention parties, both developed and developing. Article 4.1 requires that all Parties "develop, periodically update, publish and make available to the Conference of the Parties...national inventories of anthropogenic emissions by sources and removals by sinks of all greenhouse gases," and calls upon all parties to "formulate, implement, publish and regularly update national and where appropriate regional programs containing measures to mitigate climate change..."[5]

With the critical distinction that for developed countries, reports are taken with the "aim" of returning greenhouse gas emissions to 1990 levels by the end of the decade, there is not a lot of difference between developed and developing country reporting requirements as listed in the convention. The requirement to report is there for all parties. Article 12, which provides detail regarding the methods for reporting does provide for greater flexibility for developing countries in the preparation of their reports. Principally, the latitude given to non-Annex 1 parties is the time frame for submitting first communications: developing country Parties are given three years from the time of entry into force of the convention for submissions of their first communications (although least developed countries do not even have this time requirement, but are to prepare reports as soon as they can); developed country Parties are to submit their first communications within six months of entry into force of the convention.

While the guidelines developed and adopted by the INC are applicable to Annex I Parties, no similar guidance has yet been agreed for non-Annex I parties. I believe that they will closely resemble those already agreed – with critical differences relating only to the extent to which any individual developing country has the capacity or the infrastructure to provide the data. Negotiators are currently engaged in trying to develop recommendations for what ought to be included in the reports of least developed and other developing countries.

The call for developing country guidelines has been very strongly pushed by the United States, and has been strongly supported by other OECD countries. While there is a nascent interest in the development of guidelines by the developing countries themselves, to date there are no draft proposals for consideration. To date no developing countries have yet submitted communications. Although the earliest ones are not due for almost two years, the Annex I country experience suggests that nearly a year is required for the preparation of reports – and that guidelines must be in place before the preparations begin.

Speaking from personal experience, I can tell you that working on the United States communication was a considerable undertaking. The United States probably has among the best data sets in the world for presenting any kind of information. We have extensive data on demographics, we have incredible data bases maintained by the Department of Energy and its Energy Information Agency, by the Department of Commerce and its Census Bureau, by the Department of Transportation and others. We know with reasonable certainty about American's transportation habits, and how we use and develop pollution control devices. We have very, very good historical data, and there is data in the public record with regard to the development of policies, programs and measures for taking action in any area in the environment, including actions to reduce greenhouse gases. The United States in this regard is unique. Other developed countries (even countries which we think of as being heavily numerically oriented) have much more limited capacity to generate this kind of data-base.

However, even less comprehensively inventoried OECD countries have recourse to an order of magnitude more information about their populations and commercial practices than do most developing countries. For example, a well-developed country like Brazil has little historic data regarding its forestry resources, although it has recently engaged in an aggressive program to assess deforestation (and that effort is perhaps the finest of its type in the world) as a consequence of its land use management practices. The lack of data is largely a result of devoting lower priority to this issue than to other more pressing problems – that is, for Brazil, there were greater priorities than the development and the generation of statistics.

Brazil, of course, is an advanced developing country. Examining a poorer country like Kenya provides a more realistic perspective on what many countries in Africa and Asia might expect to be able to produce through their communications efforts. While Kenya is one of the most developed countries in East Africa, and it has very limited resources to devote to an examination of how many automobiles are currently on the road. In fact, there is no Kenyan database that targets that particular piece of information. How is Kenya to generate statistics which would allow them to inventory their current emissions? Even more difficult, how does Kenya project the effects of efforts to reduce those emissions over the next five, or ten, or twenty years? These are the kinds of problems with which the international community is now grappling as it seeks to develop guidelines reporting for developing countries.

There are some emerging solutions to these kinds of problems. The Intergovernmental Panel on Climate Change (IPCC), while best known for some of its efforts in the scientific assessment of climate change, has devoted substantial resources to looking at techniques for inventorying greenhouse gas emissions. Their conclusions provide methods to extrapolate from the little data that is available to what the total picture might be (admittedly with substantial uncertainties). To do this, the IPCC relies on baseline factors measured in selected communities, and emissions factors derived from small samples in one country that may be extrapolated to other countries. The final results should at least provide a first estimation of the magnitude of national emissions, and assist in prioritizing actions to mitigate those emissions.

There are also several programs underway within the international community to assist developing countries in the preparation of their national communications. The United

States, through its country studies program, which is a $25 million effort to help developing countries and countries with economies in transition to market economies develop basic information that could lead to reporting and to taking actions, is one example. The United Nations Environment Program (UNEP), has put resources into reporting programs, and the United Nations Development Program (UNDP), and the Global Environment Facility are supporting similar efforts. Notwithstanding these programs, improved guidance for developing country reports will be a critical component of the next steps under the Convention process, and one that can be expected to have high priority over the next year.

NEXT STEPS FOR REPORTS

Let me turn now to the issue of next steps — particularly for OECD countries — with respect to development of reporting guidelines, the preparation of communications by Parties, and the review of these communications by the FCCC and its subsidiary bodies. At the moment the guidelines adopted for reports mirror the timeline established in the FCCC: Parties are to estimate the effects of measures to the year 2000. That's only now about 5 years away, and some longer term estimation is clearly going to be required no matter what regime is ultimately chosen. Reporting should continue to track any new mandate and guidelines will need to be reset to follow any new agreement.

One of the most difficult problems that countries are likely to face in developing their reports is that of projecting measures. At present, while the guidelines suggest that countries might include in their reports a longer-term projection of the effects of their measures on national emissions, very few countries have chosen to do so in their first communications (the United States was one of these few; our own plan suggests that without further future measures, U.S. emissions will rise between 2000 and 2010). Techniques and methodologies for long term projections are rudimentary. Even the most optimistic modeler cannot claim to fully believe his or her projections. And the error is magnified the further into the future projections are made. Notwithstanding this difficulty, the need to evaluate progress and to undertake the most effective and least costly measures will require such projections. While the current reporting guidelines are only suggestive of the need for linger term projections, even this "soft" recommendation was strongly opposed by many in the OECD. Overcoming this concern, and setting agreed standards for the development of longer term projections is therefore an area on which the Parties must focus their attention.

A second area which the Parties must provide further guidance is on reporting standards for individual measures. While the FCCC itself calls for parties to provide "a detailed description of the policies and measures it has adopted to implement its commitments" and "a specific estimate of the effects that the policies and measures...will have on anthropogenic emissions by its sources and removals by its sinks,"[6] the current guidelines do not provide a format for the presentation of such information. Inasmuch as many of the ideas surrounding next steps relate to undertaking common measures or policies, there is a clear need to evaluate — in a common manner — the current set of actions being taken by countries.

In other words, setting a goal may be a first step, but ultimately, it will be the actions taken to meet the goals that will be the acid test of success in combating the climate change problem. Of course, there is a catch: until the international community has more accurately and precisely defined precisely what actions should be taken, enumerating reporting requirements may be premature – or focus on the wrong actions.

Because the FCCC does not mandate a particular set of policies and measures, countries have been free to choose those most closely applicable to their national circumstances. However, this has presented a problem for the international community in comparing the effectiveness of measures, and evaluating the aggregate consequences of national efforts. Clearly, elaborating on this aspect of reporting would be an appropriate avenue for consideration by the Parties as they revise the reporting guidelines.

A third area critical area for enhancing reporting guidelines is that of reporting on joint implementation. Joint implementation is a concept, embedded in the FCCC, in which one country (or a company within a country) can claim domestic "credit" for actions taken to reduce greenhouse gas emissions in another country. While there is no clear consensus at this time as to how large a percentage of global emissions might be reduced through joint implementation activities, it is equally clear that no guidance has been provided by the international community with respect to the reporting of these actions.

Clearly, reporting on joint implementation cannot only be a reporting requirement only for the country sponsoring the activity; because it involves more than one Party, the reporting guidance must detail methodologies for both host and sponsoring countries. To date, there are no methods yet developed that offer guidance. And, one of the problems that many JI skeptics see is specifically tied to this lack of reporting guidance, arguing that without guidelines for developing countries, double counting of credits, cheating on efforts, and lack of transparency will be rampant.

While there are likely many other aspects to the future development of reporting standards which will take considerable work and lengthy negotiations (e.g., development of standards for inventorying the less common greenhouse gases, decisions with regard to reporting on bunker fuels, etc.), the areas discussed above will be among the most important. Participation of groups like this one will clearly be influential in assisting with developing sound practices, and will need o be regular and active participants in this process.

THE NEGOTIATIONS: STATUS AND NEXT STEPS

While I have alluded to many issues related to next steps under the FCCC, let me turn briefly to an explicit discussion of where the negotiations currently stand, and what might be expected of them over the next several years.

First, where are the negotiations now? Countries have just concluded their tenth negotiating session (INC-10); there is a final preparatory session (INC-11) scheduled prior to the first meeting of the Conference of the Parties, which is scheduled from the 28th of March through the 7th of April in Berlin. The last three days of the Berlin session are to be a high-level session - at the ministerial level. It's not yet clear who from out of the United States or other countries will attend that session. We do know that Germany intends to open the ministerial segment of the meeting with a statement by Chancellor Kohl, offering the international community a perspective on the importance which Germany attaches to this issue and to this conference.

There are a number of actions that the Convention mandates, or that Parties are expected to take at their first session this spring. Parties are expected to adopt recommendations with regard to reporting (including some of those discussed in detail above). They are expected to adopt decisions on financial matters, including decisions with regard to the linkages between the Convention and the Global Environment Facility (GEF). They are expected to adopt decisions with regard to the rules of procedure for the operation of the climate change convention, and with regard to the work programs, meeting schedules and Bureaus of the FCCC's subsidiary bodies. They are supposed to adopt or make recommendations with regard to criteria on joint implementation (although on this matter, the Convention itself only requires that the Parties "take decisions regarding criteria for joint implementation,"[7] which could mean postponing them, delaying them, or taking a final decision to adopt).

The Parties are also supposed to review the adequacy of the commitments in Article 4.2 of the convention. While the Convention does not specify what exactly is to be done with the results of the review, the Parties are given some guidance: "[b]ased on this review, the Conference of the Parties shall take appropriate action, which may include the adoption of amendments to the commitments…"[8] There are currently three proposals that have been tabled with respect to next steps on the issue of adequacy. The first proposal is from the Alliance of Small Island States (AOSIS). It calls for a 20-percent reduction in carbon dioxide emissions by the year 2005 from 1990 levels for Annex 1 Parties. It also calls for full, preferential transfer of necessary technologies for developing countries to enable them to move forward in their efforts to address climate change. While this proposal will in all likelihood be formally considered by the Parties at Berlin, I do not believe it will be adopted there — it is too far from the mainstream of what countries currently find politically acceptable.

The second proposal to be tabled is from Germany. While this is essentially proposed as an amendment to the AOSIS protocol, it does not necessarily indicate that Germany would be willing to fully endorse the AOSIS proposal in all its particulars. The German proposal (which was not tabled in either the form or timeframe to require it to be formally considered by the Parties in Berlin) largely focuses on the area of specific measures. The German proposal also elaborates on one aspect of the AOSIS proposal which calls for a new body under the convention process to look at specific measures and how they might be implemented in next steps. In their recommendations, the Germans list various sectors and various kinds of measures that might be considered by Parties for future action. However, the list has an all-inclusive character which essentially renders it less useful than a more targeted approach might be. By listing almost every option, it targets none – and is little different than an open invitation to parties to consider additional, unspecified and unprioritized future action. While the efficacy of specific measures are ultimately the test of success, it seems that a narrower focus will be required to determine in which direction to look for those steps.

Another proposal has been tabled by the Swiss government. Similar to the German proposal in some respect, it also targets measures. The Swiss propose to establish a panel to review and evaluate measures, specifically seeking to find commonality between actions that countries are or could be taking, with the ultimate goal of adopting specific harmonized measures. The Swiss perspective is driven by concerns over competitiveness — if all countries take a similar (or identical) action, the competitiveness concerns can be blunted.

These and other proposals will be on the table, and both formally and informally discussed by delegates at both INC 11 and at the Berlin meeting. However, notwithstanding the urgency that representatives of these countries bring to this issue, I do not believe the Parties, in Berlin or before, will adopt any or even part of any of these proposals. This does not mean that these proposals will be buried — only that prior to Berlin, there will be inadequate time for their adequate consideration and possible approval.

What then will the Parties undertake as next steps under the FCCC? To answer this question, or at least to speculate on an answer, I think it is critical to first examine the rationale behind taking action at all. I find myself persuaded by technical arguments (many of which have been well summarized by the IPCC) that ultimately continued emissions of greenhouse gases will lead to climate change. While the local impacts may be uncertain, the changes are likely to be disruptive. Unfortunately, it is also difficult to look into the future to assess how effective specific actions might be to alleviate the problem – either through mitigating the change, or through adapting to the new climate system. At this time, decisions must be made in the face of considerable uncertainty, both with respect to the impacts and consequences of climate change, and to the costs of taking responsive measures. Perhaps even more unfortunate is that detailed answers to these questions are not likely in the near term.

In the face of this uncertainty, and also, in the face of the lack of international political will for more aggressive action, the options are relatively circumscribed. Perhaps one of the clearest enunciations — and in my mind most useful — regarding the emphasis that ought to be taken was provided by David Fisk, the chief scientist for the UK environment agency. He said that the international community needs to retain a degree of "dynamic tension in the system", so that things could move quickly if and when a public willingness and agreement to undertake action becomes apparent.

At this time, the political system globally (and certainly in the United States), is not willing to undertake radical new steps to address climate change. At the same time, the climate issue continues to raise concerns. It has not disappeared from the table, nor does it seem likely to prove a chimerical or ephemeral problem. The level of willingness on the part of U.S. industry to adopt voluntary measures has not substantially slackened since the elections last month - in my mind an indication that corporate America does not believe the problem will go away if we ignore it. Instead, industry is saying that long-term solutions have to be carefully evaluated, and that costs and benefits of taking actions must be considered, but at the same time acknowledging that some kind of actions will need to be taken.

To what actions can the international commmunity now agree? For the near term, one of the easiest is likely to be revised reporting standards as discussed above. Specific international measures may also be a near term option if measures can be found that meet with broad approval. Options for consideration include research and climate monitoring standards, as well as data and information exchange.

In the longer term there are some potentially significant problems which could act to prevent international agreement. For example, there are competitiveness issues related to agreement on harmonized instruments (e.g., international taxes or standards). Issues of intellectual property and property rights begin to impinge on matters of transfer of technology. Lack of substantial capital in the donor countries inhibits large financial transfers to developing countries to assist them in taking actions.

However, some options do exist for taking next steps. Promoting joint implementation, technology development and a developing a common menu of actions from which each country may choose may all be fruitfully explored. While an aggressive approach such as that proposed by the AOSIS countries may not see action until the scientific certainty surrounding climate change is greater, a more limited agenda for action is still possible.

What will it take for a more aggressive next step? I am not sure that scientific certainty alone would drive immediate adequate – it will take some time for public awareness to catch up, and ultimately it is public opinion that drives political movements (albeit shaped by the science). However, even in the absence of scientific certainty, catastrophic changes in the climate system could drive action. Three or four years of successive droughts (like that of the US. in 1988), or an equivalent number of successive years of Mississippi flooding (like the flood of 93/94) will drive public opinion. Other non-climate drivers also enter into force: if the insurance industry starts pushing up rates in the belief that the climate is changing – we would not need to be certain about the change to feel the impact of the increases. If new technologies emerge that have ancillary climate change benefits, I am sure that their developers will argue these benefits in their sales pitch.

What kind of actions could be ultimately taken? In the longer term, targets and timetables for reductions of greenhouse gases are probably an option. Leading up to these, however, will be the requirement that countries develop measures to meet ambitious goals. For targets to be realized, countries must first aggressively promote technologies which are currently at the edge of commercial feasibility - but which might yield dramatic reductions in emissions: options in the energy sector such as solar photo-voltaics, wind power and biomass; options in the transport sector, including examination of new infrastructure; options in non-CO_2 gases, such as techniques to reduce methane and nitrous oxide emissions.

It remains to be seen whether the problem of climate change warrants this kind of approach. The negotiations leading up to, and in Berlin, will determine, at least the near term outlook for the international debate.

ENDNOTES

1. Annex I Parties include: Australia, Austria, Belarus, Belgium, Bulgaria, Canada, Czech Republic, Denmark, European Union, Estonia, Finland, France, Germany, Greece, Hungary, Iceland, Ireland, Italy, Japan, Latvia, Lithuania, Luxembourg, Netherlands, New Zealand, Norway, Poland, Portugal, Romania, Russian Federation, Spain, Sweden, Switzerland, Turkey, Ukraine, United Kingdom, and the United States.
2. The following countries had submitted reports as of January 1, 1995: Australia, Austria, Canada, Czech Republic, Denmark, Germany, Hungary, Ireland, Japan, Netherlands, New Zealand, Norway, Spain, Sweden, Switzerland, United Kingdom, and the United States.
3. Countries which had ratified but not submitted reports by January 1, 1995 included: the European Union, France (which has since provided its report), Iceland, and Italy.
4. Daniel A. Reifsnyder is currently the Director of the Office of Global Change at the Department of State, and has been alternate head of delegation for the United States at the negotiations leading up to the first meeting of the Conference of the Parties in Berlin.
5. FCCC Article 4.1(a) and (b).
6. See FCCC Article 12, paragraph 2 (a) and (b).
7. FCCC, Article 4.2(d).
8. Ibid.

41

DISCUSSION WITH JONATHAN PERSHING AND ELIZABETH THORNDIKE

Ms. Werner: Yesterday the UK indicated that they would be the first national action plan review starting in January. Do you want to comment on that?

Dr. Pershing: Yes. There are two things to note about that. It's either the U.S. or the UK, I don't think it's been finally determined yet. The second thing is that there was a substantial question with regard to how the in-depth review should take place. The debate was over whether or not it was appropriate to have what are called country visits. In other words, could the reviewing team actually go in person and speak with representatives of the government in the country that had prepared the plan. The only countries that strongly endorsed this were the UK and the United States. And both countries have requested and volunteered to have their plans reviewed through this in-depth process.

Objections to having this review process stemmed from two issues. One, I think a fear by countries that they may, and this is perhaps unfair, but may uncover things that they wanted to keep quiet. The second is a sense that it was unnecessary, was going to be costly and really more difficult logistically to arrange to have these country visits, that a paper review process would be more than adequate to meet any kind of review requirements.

So it will be either the U.S. or the UK. Currently those are the only two countries that have volunteered to have their reports reviewed.

Ms. Thorndike: I want to take the opportunity to share with all of you a rather interesting development which I think speaks to the role of this issue in the short term. Not the long term because I think we have all come to recognize, and everyone in this room knows the global climate change issue is a very long-term issue, and we do not tend to focus very well on long-term issues.

But let me suggest something very short term which may be a worthy indicator of where we're headed in the coming year. The Center for Environmental Information, as you know, has put on a conference annually for about ten years, the recent five or six on this particular topic, here in Washington. We put on lots of conferences and programs on all sorts of subjects and we have is a good records of how the flow of registrations go with conferences.

The early returns and the mid-stream returns gave every indication that this conference would have the largest number of participants of any that we've ever had. Certainly the enthusiastic steering committee who put this together felt that way in dealing with the National Climate Action Plans and the commitments to the Convention.

Following the 1994 election something very dramatic happened to the registrations for this conference. They dropped off remarkably. We have no parallel in any of these conferences . I point this out to all of you because it may be an indicator, not necessarily of what is or what will or what should be happening, but certainly of a certain perception out there of this issue in view of the election results. And as you know these conferences are attended by people from all around the country.

So I'm talking about what happened in states and statehouses as much as what has happened in the Congress.

42

CLOSING REMARKS

William Nitze

United States Environmental Protection Agency
401 M St., SW
Washington, DC 20460

I want to thank the Center for Environmental Information for staying the course with the climate change issue. This is still a great issue. It is an issue of tremendous importance for the 21st Century, and it is an issue which is not going to go away.

And furthermore, I'd like to sound a note of optimism. I think it is an issue which can be addressed by the United States in a manner which commands bipartisan support.

My departure from the State Department was initiated by my work on precisely this issue. But what I want to start my remarks with today is to pick up on the work of my successor as the Deputy Assistant Secretary for Environment at State, Bob Reinstein. He got the funding for and introduced into the Climate Change Convention the key concept of national reports. And to my knowledge the $25 million which Bob Reinstein got the Bush Administration to allocate is still the major source of outside funding for efforts by developing countries and countries with economies in transition to actually figure out what greenhouse gases they're putting in the atmosphere, what the sources of those gasses might be, what sinks they might have for absorbing those gases, what mitigation options might be available and how they might go about actually exploiting those options.

So I'm going to lay out for you the Country Studies Program, funded by Bob's $25 million, which was created just prior to the time that the convention was signed in Rio, a program that I think we can all be proud of and that's made considerable progress.

I will then go on a to talk about some of the ideas that we are considering in EPA and elsewhere in the administration for continuing this program in a slightly different guise.

Then I will spend a few moments on some examples of EPA programs abroad which would fit in with this extended strategy. And finally, I will throw out some slightly more farfetched ideas as to how we might proceed before and after the first conference of the parties in Berlin in late March of 1995.

This program, which I'm sure many of you are familiar with, is based on a well developed foundation of interagency and international donor cooperation. We are working with other countries to help fund these efforts in developing countries and Eastern European countries. And we have a number of mechanisms here in Washington to manage the program and make sure that there is proper coordination.

There's an Interagency Coordinating Committee chaired by the Department of State, there's a Country Studies Management Team with staff seconded from EPA, from DOE, from

Evaluating Climate Change Action Plans, edited by White
Plenum Press, New York, 1996

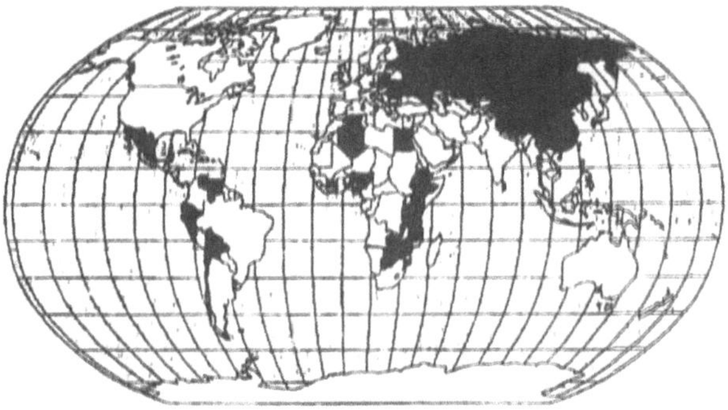

Figure 1. Program is supporting studies by 53 countries in all regions.

NOAA, from USDA and from the Department of State. And then we work with a number of other donor countries and international organizations, including UNEP, UNDP, the Global Environment Facility, IGBP Start, the Intergovernmental Panel on Climate Change, and OECD as well as a number of our leading OECD partners; for example, Germany, the Netherlands and the United Kingdom.

At the moment the program is supporting studies by 53 countries in all regions. There are 26 countries that received support in round one, studies that started in October of 1993, and there are an additional 27 countries receiving support in round 2, studies that started on October 1, 1994.

The countries receiving U.S. support under the countries studies program have for the most part the greatest greenhouse gas emissions and the most rapidly growing populations. We have had an opportunity through this program to actually influence policy-making on this issue in many, if not most, of the key countries around the world.

I will just mention one in passing, and that is China. We have a huge opportunity to further our work with the State Planning Commission and other agencies, both at the central, provincial and municipal levels in China in looking at greenhouse gas emissions and in devising mitigation strategies. And this opportunity was really created as part of this $25 million initiative.

The 53 countries listed in figure 1 represent a substantial share of the world's carbon emissions, an substantial share of the world's methane emissions, and an even more substantial share of the world's projected population growth.

We believe that the program has built support around the world for the Framework Convention on Climate Change. And there are some reasons why. The results are forming the basis for INC communications. They have provided badly needed enhancement to U.S. stature and leadership in the negotiations. The technical support that we have provided under this program is becoming an international standard.

The individual work we are doing in these countries is really the foundation for the current comprehensive international assessments. And the partnerships that we are creating

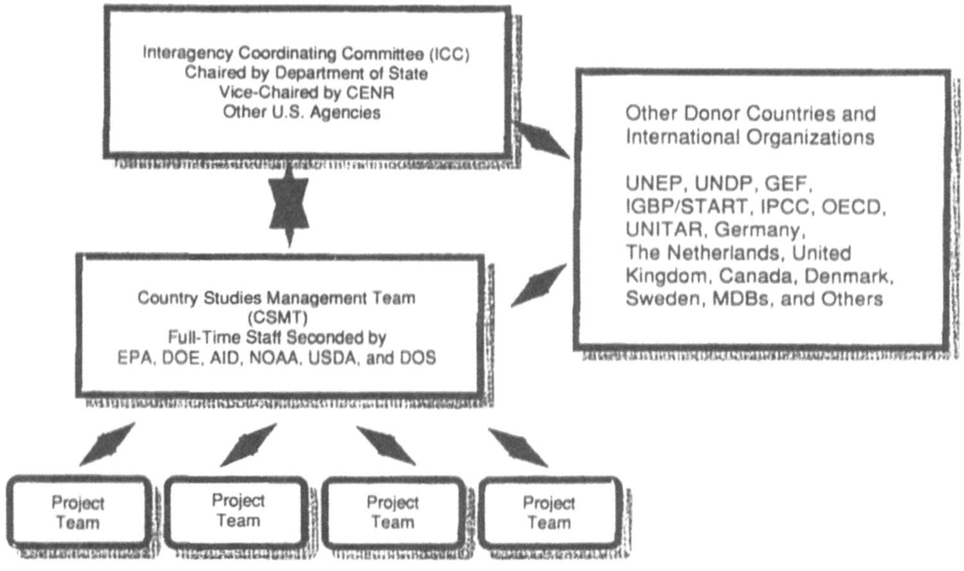

Figure 2. Interagency and international donor cooperation is basis for U.S. country studies program success.

are the basis for future cooperation, negotiation, technology exports and a more technology-driven climate strategy.

This figure gives you graphically what I described early a moment ago, and that is the interagency and international donor cooperation mechanism, the Interagency Coordinating Committee and the project teams grouped under the Countries Studies Management Team. And then in the big box, the links to the other donor countries and their national organizations.

The key factor that distinguishes the U.S. Countries Studies Program from programs provided by other donors is technical assistance. It obviously is not enough to simply provide funding for individual developing countries or Eastern European countries to do their reports. You have to get in there with them and in a hands-on way provide technical support in developing baseline emissions profiles and mitigation options. This helps to build a network of people throughout the world who are thinking and talking in roughly the same way about how these problem can be attacked.

The scope of the technical assistance provided for the studies is quite comprehensive. It covers training, including many, many international workshops. It includes tools, particularly guidance documents, models and data sets, and it includes technical cooperation and partnerships. We think if we can continue this process and strengthen it, it will promote sustainable development.

Table 1 shows you the difference between what you've been reading in your World Resources Institute book and what the Countries Studies Program work came up with. And in some cases, particularly in the case of the Czech Republic and Ethiopia, the differences are quite dramatic. It turns out that actual CO_2 emissions based on the analysis conducted under the Countries Studies Program is significantly lower than had been published in the WRI report, there are also discrepancies on methane, but they go in different directions. This illustrates the importance of careful comparable work to establish baseline emissions inventories and how far away we are from an accurate and comprehensive knowledge of those inventories around the world.

Table 1. Assisting countries in meeting FCCC inventory reporting: U.S. country studies improving global emission data

	CO_2 (MMT)	CH_4 (MMT)
Czech Republic (USCSP)	161	.834
Czechoslovakia (WRI)*	191	.710
Ethiopia (USCSP)	2.10	.945
Ethiopia (WRI)	2.82	1.400
Peru (USCSP)	19.7	.429
Peru (WRI)	19.2	.400

*WRI figures are for 1991 since no national numbers for 1990 are available.

In many cases the work done under the countries studies program is the first attempt by countries to estimate emissions. It has provided more comprehensive and accurate estimates than any other prior work. The estimates have been endorsed by national governments, and of course they provide the basis for mitigation assessment and developing ongoing research programs.

In another program, we look at the whole question of vulnerability as part of these comprehensive assessments. Some examples of vulnerabilities that we have identified and where the host countries are becoming more aware of the risks are the water availability in Zimbabwe, potato productivity in Russia, runoff from the Elbe River and flooding of low-lying coastal areas in Peru.

There are many, many other such examples. And valuable work is being done by the Climate Institute and other organizations in coming up with additional information of this kind about vulnerabilities to climate change and the basis for adaptation strategies. But this is a very important dimension to our program, not only because of the importance of the adaptation issue per se but because of the importance of building a political consensus behind meaningful action.

This process is a necessary first step in determining what really are the cost-effective greenhouse gas mitigation options in countries receiving U.S. support. And sometimes you get rather surprising answers. We at EPA have found, for example, that coal bed methane is in many countries a much bigger opportunity than we had thought even a few years ago and as we continue our work, we expect to learn more new things and identify more possibilities for mitigation policies and projects.

Though this program is not munificently funded — $25 million is not a huge sum of money — and if we were to add $5 million or $10 million or $15 million to that, that also would not be a huge sum of money.

But the tremendous leverage in this approach is in forcing the countries who have volunteered and asked for assistance to organize themselves so they have the capability to come up with accurate inventories, come up with a preliminary plans and, most importantly, a process for revising and adding to that plan over time. And without that institutional capability there will never be sufficient confidence to agree to a more far-reaching international strategy.

We do have broad interagency support for continuing this program, more than 30 countries have requested ongoing support, including some of the major greenhouse emitters and that in a sense is a unique U.S. initiative which sets the stage for the next phase. We hope to see it lead to the establishment of pilot projects and assisting countries with the identification of technology and financing needs and resources.

Now, let me move to a more free-form discussion as to where this process may go in the future. We face a very interesting challenge in Berlin. There is an emerging consensus among the OECD countries that we have to follow-up on the last INC meetings and establish a new stabilization goal. The goal established in the Framework Convention is not sufficient and we need something that will lead to greater and more rapid reductions in emissions so as to produce lower concentrations into the next century.

The difficulty with going into the Berlin conference with the sole idea of somehow starting a negotiating process towards a new goal is that we're in a political environment there. In many other countries of the world it's not going to be easy to get support for further reduction measures unless, and this is a critical condition, unless there is a sense that those measures will lead to enhanced economic growth, quality of life, job creation and other goals which respond to the perceived political needs of the moment.

Whereas in the past much of the concern about not sacrificing economic development was concentrated in the developing countries and to a certain extent in Eastern Europe, now it is also appearing in the OECD itself. There is tremendous concern, which was reflected in this election, in what is happening to the U.S. economy as a result of the overall globalization of the world economy.

There is an even greater concern in Europe about not only what has happened but what is likely to happen as Europe is forced to adapt to the new global economy. Even in the richer countries of Asia you are seeing concerns about losses of employment, hollowing out and so forth as a result of competition from very ambitious and lower cost neighbors.

In this environment we have to devise an approach to this issue which links our climate change strategy with broader economic development and social development goals. And we have to minimize the reality or perception of conflict. And it seems to me that the way we have to do that is to focus on what I will call the technology-oriented approach.

We need to find ways of accelerating the deployment of low-emitting technologies, technologies that enhance sinks or recovery in a manner where the countries in which those technologies are being deployed perceive them as helping their more rapid modernization and simultaneously solving economic, social and environmental problems.

I think this is eminently doable, but it is going to take a new approach and it's going to take considerable work. Let me throw out an example of a conventional technology which we are working on very hard in the United States, and which could be tried perhaps in a slightly different way in other countries. We have made a political decision in this country to work towards a green car.

Now, this is a continuing evolutionary process. We've made considerable headway since the 1970s but under the Clean Air Act Amendments of 1990 and the Ozone Transport Commission decision and the spread of the California rules with respect to low and zero emission vehicles, we're going to have to accelerate that process. EPA will very soon have an important and positive announcement to make with respect to the New England states.

We are undertaking this initiative in a context where we're relying very heavily on regulation and in my view not enough on pricing signals. We all know the reasons for that, the history of the BTU tax and so forth.

As we look at rapidly growing automotive markets such as China and India and some of the key Latin American countries, what can we do to make sure that, as more and more cars get on the road in those countries, will there be intelligent transportation planning, emphasis on public transportation,and land use planning ?But even with all of those things, and even with a high cost structure and high taxes, you are going to see a rapidly rising use of automobiles throughout the developing world.

So it becomes very, very important to make improvements in the environmental compatibility of those vehicles. Now, how might we go about that? I happened to be meeting with the association that represents the non-U.S. automobile manufacturers in the U.S., and we batted around some ideas. And one thought might be to take the golden carrot approach that we've developed for refrigerators in this country and apply it on a larger scale with respect to vehicles.

If, for example, China provided refunds that would give a $500 or $1000 a car subsidy to vehicles that were the result of winning a competition for who could produce the greenest car, you probably could get a very lively competition involving all of the world's major auto manufacturers and a vehicle that went considerably beyond current vehicles in efficiency, low emissions and other environmentally desirable characteristics.

Now, that's just one idea. It's not something the U.S. could promote on its own. It would probably require international financing of some kind. There would be quite a lot of negotiation that led up to it. But I think it's an idea worthy of study. I think we do have to develop mechanisms to take technologies that are already available and make sure that the developing countries can actually leap frog into deploying those technologies sooner rather than later.

Now, there's a second leg, and that is the transition to really non-emitting technologies, non-fossil fuel or very low emitting fossil fuel technology, what some might call the hydrogen economy of the future, or the solar-hydrogen economy of the future. We have to think much more creatively about engaging major developing countries in a joint research and development effort where they would work with our best people in developing products for their markets and for that matter for the global market.

There are concerns about competitiveness and trade implications of this strategy. My answer would be that the developing countries are where the economic growth of the future is going to be. Any progress with such an initiative at this time would lead to tremendous investment opportunities for OECD firms, particularly U.S. firms. It would in a sense accelerate growth of the world economy, but most importantly in a way that minimizes the environmental risk from the increases in greenhouse gas emissions and global warming.

And, as you all know, there's a nice fit between technologies that minimize greenhouse gas emissions and technologies which reduce particulates, ozone and other sources of pollution of more direct local impact.

Any greenhouse gas mitigation strategy, in almost any country I can think of, is going to have to be sold on the basis of its shorter-term benefits in addressing environmental concerns of citizens living in those countries.

The reason that the South Coast Air Quality Management District has been able to be quite draconian with respect to automobile standards is because the people who live in Los Angeles are very concerned about dirty air in terms of their health and their quality of life.

Similarly, greenhouse gas mitigation strategies in other countries are going to have to be sold not in terms primarily of their greenhouse gas mitigation effects, although obviously those are important, but largely in terms of their more immediate benefits for human health and quality of life. And in my view the key to doing that is technology diffusion, and the key to achieving better technology diffusion is getting a better incentive structure. And that is going to require a lot of creative collaboration between the OECD and the developing countries. I think it's the only way to go.

43

DISCUSSION WITH M. GRANGER MORGAN, WILLIAM NITZE, AND JONATHAN PERSHING

Ms. Werner: The U.S. plan says that measures will meet stabilization and you say the U.S. is the only country to do so. But if the reality is that the U.S. has a gap, no matter how large, is this not in a sense false advertising?

Dr. Pershing: Actually the U.S. plan is not the only plan to propose stabilization target. The United Kingdom says that its plan will do the same thing and the Dutch say their plan will do the same thing. What I tried to say was that none of the states met the guideline requirements for the communication. And that is different from saying that the plan would in fact meet stabilization commitments.

The stabilization commitments that were made are unilateral commitments. They are not commitments that are embodied in the Climate Convention. They are commitments that were made by the President in a speech that he made on Earth Day.

The current expectations in the plan, as drafted, is that, with the existing economic situation at the time the plan was created and with funding of the program as the Congress at that point looked like it might fund, we would have made stabilization.

We also make a note, and there are two parts to the plan. In the very end of the plan there's a chapter on the future in which we discuss the kinds of factors that could influence our ability with the existing measures to meet this stabilization target. We do not at this time have a final revised set of numbers to present publicly.

Our estimate is that because the plan relied on funding which has not yet been available, because the plan used GNP estimates and projections which are not currently what we are seeing, including estimates of the oil price, we will probably have to seek additional measures to meet or return to 1990 levels by the year 2000.

Ms. Werner: Why are governments so reluctant to clearly and candidly acknowledge the growing gaps between stated goals and likely outcome? Is there a point where that reluctance to acknowledge those gaps can undermine international and national efforts to limit emissions?

Dr. Pershing: My sense is that there are a couple of things operating here. I think the first thing is a degree of embarrassment. I think the countries that have made the strongest

And the larger your apparent gap, the more difficult it is both to maintain that front and, much more important for many countries, the more difficult it is to call for stringent next steps. Therefore, I would argue, the quieter you can keep your gap the easier it is for you to maintain that next steps ought to be taken.

But you are unlikely to be able to maintain that fiction for very long. I think that the review process already is beginning to demonstrate the kinds of problems that countries are having in taking the measures they've announced or meeting the goals that have been proposed for those measures. And I believe that next steps are unlikely to be taken in the absence of some kind of relatively transparent view of what countries can in fact achieve.

Ms. Werner: The U.S. provided estimates of reduction from individual measures, but you indicated other OECD countries did not. Why? Do OECD countries not have data available and simply choose not to provide that information or are they unable to quantify them? And how can the U.S. best push or persuade OECD countries to include more data? What's your plan?

Dr. Pershing: That's a really interesting question. There are a number of pieces to that. Let me provide some anecdotal experience. In the course of the development by the Annex 1 parties of the guidelines for communications we looked at the communications that were available in draft form. At that stage there were only three.

The United States had presented a communication to meet President Bush's recommendations. That was done in 1992. The Germans had done a draft plan and the Canadians had done a draft plan. A team process was developed to visit countries that had done plans and countries that were currently in the process of developing their own communications.

I was one of the representatives on the OECD team to look at the German communication. Germany does not have a federal government approved projection of GNP. They do not have a government sense of the future.

They have an Economic Ministry, and a Finance Ministry that both look at numbers, but there is no formal agreement as to what the situation really is. Instead they rely on private commission numbers for their public data, and as a consequence there's a degree of deniability established by having a process which is not a public forecast. This kind of things propagates itself through much more of the German system. So you don't have the public numbers of what GNP growth is going to be.

Germany does not provide those kinds of numbers and the United States is not likely, through any kind of pressure we can apply to Germany, to call them on having the new numbers. At the same time, we can show that this failure makes it much more difficult for the community to move forward with some kind of global assessment.

I believe that the Germans in the next iteration are likely to try using these more informal sources so that they can include in their plan some set of suitable caveats, saying that these do not represent formal government-approved figures or text.

The same kind of thing can be done in the Netherlands, or in the UK. I think that a number of individual communications do provide estimates of specific measures. The second part of the answer is looking at an integrated model to examine the overlap between measures, a critical part of doing an analysis.

For example, supposing you have a measure which called upon countries to consider more efficient light bulbs like compact fluorescent lamps. At the same time you have a measure which calls upon the building community to increase the efficiency of buildings. There is a clear overlap between those two sets of measures. The building community may in fact be installing compact fluorescents as part of its program to make more efficient buildings.

How much of the effort is a consequence of the building measure and how much of the effort is a consequence of the CFL measure? It's very, very difficult to isolate one and distinguish it from the other. Programs which do these kinds of analyses are also very difficult to create. I think that Dr. Morgan could speak to the difficulties in modeling these kinds of analyses with some level of detail.

There are a number of places where the U.S. can contribute to the process, and I think that ultimately we will converge on a much broader degree of estimates and projections for future emissions.

Ms. Werner: In a democracy most policy-making is driven by public opinion. Since public opinion rarely crystallizes in the absence of a perceived crisis, and even then often shifts dramatically, it's probable that policy proposals developed outside of a crisis context will meet significant public resistance in the early years and prove inadequate in later years. How can analysis models take natural human shortsightedness into account?

Dr. Morgan: Well, let me start by saying a word or two about the way that I think analysis shapes policy. My model of process is not that I or others as analysts sit here, do a piece of analysis, hand it to the assistant administrator at the moment he has a problem and he says, "I guess Morgan's right," and he goes off and does it.

Occasionally that kind of analysis occurs, but the way the kinds of analyses I was talking about are realized is that gradually a community begins to reach similar conclusions about how a problem ought to be framed and thought about. We and the group at Batelle Northwest Labs and the group at MIT, each begin to discover: "gee, the conclusions I reach look real different when I put aerosols in or I leave aerosols out."

In the future we need to worry that as you promote this high energy efficiency technology you are also promoting rapid sulfate aerosol production. And in the short term that's likely to produce more rather than less warming. Factor that into your thinking.

So that's the first comment, that the model is not one of tight coupling but is more diffused. The second statement, though, is that there are at least some strategies which can pursue both the long-term climate objectives and a variety of short-term immediate objectives.

And to the extent that these can be pushed, that the general public begins to understand the climate problem better, that the sorts of issues that I talked about are addressed, that people have a slightly more focused mental model of what matters in the climate issue, we have some chance of making progress.

But if you're asking me: how do you get the United States population or any population to take very large short-term financial hits in the interest of solving a problem whose signal we may not see for 40 years yet, my reaction is good luck. I don't think you're going to.

Ms. Werner: Dr. Morgan characterizes integrated assessment as a method of capturing distributed decision-making where a great many individual decisions will account for a problem. Where does integrated assessment account for changes in national decision-maker behavior due to the process of international negotiations itself? Also previous INC Framework Convention negotiations suggest that the bargaining process may change risk perception and bargaining stakes.

The second part. Also you suggest that SRI is a more effective predictive tool for international negotiations than GWP because the latter is too simple. Doesn't a more complex scenario-based index make international verification in confidence building more rather than less difficult?

Dr. Morgan: We had a state legislature in this country some long time ago that decided that having pi equal 3.14159 was a royal pain in the neck, and they literally passed a state law that pi would be 3. And it made life simpler, but it was also wrong. It didn't work.

The problem with GWP is that it says that a cost incurred today, an impact occurred today, will be valued in the same way as an impact incurred in 150 years. Most of us don't think about the world that way, appropriately. Our sense is that there are going to be hundreds or millions of decision-makers spread all around the world, who are going to make not single decisions but sequential decisions.

So we are groping for ways to figure out how to incorporate that attribute of the problem in integrated assessment. We now have this large simulation, the ICAM II model. One of the things we're doing at the moment is building independent agents, little computer characters, who have decision roles. You start running a model, and the China agent comes along here and starts making its decision, and the OECD agent comes along here and starts making its decision. How does the world evolve with those agents and those decision roles? All right. Now let's start over again. We'll put in different agents who have different decision roles.

In other words, it's a vehicle that allows you to explore the structure that the various distributed decision-makers face as the world evolves. That's about the best you can do. I mean, if you're asking me to build a model in which I provide predictive evidence about how international negotiations are going to unfold, I'm not going to do that.

Dr. Pershing: If one were to use, say, a slightly more complicated system and instead of taking a single GWP, look instead at a series and make some risk assessments on the basis of different time horizons, that might be an intermediate acceptable way to substantially simplify what would otherwise be a scenario-based model.

The reason I ask that, considering the political difficulty that working group three of the IPCC has had in having any acceptance of any scenario for the future, trying to generate political acceptance of what will ultimately become a potentially very important issue like GWPs on the basis of those scenarios, would indicate that you cannot do the scenario-based modeling in a political field, probably not in decades.

Dr. Morgan: Well, it depends, of course, on how sensitive it is to the scenario. I mean, if you can produce a wide range of scenarios and show that the scenario-based assessment produces, say, methane indices that are factors of three or four or bigger than GWPs, then it probably doesn't matter too much which scenario you choose. You could probably agree that an index for methane that's twice the GWP index is the sort of ballpark we ought to be in.

And I would not for a moment argue that the right way to go about this is an exhaustive analysis where everybody agrees on a scenario and then uses the numbers that come out of it as thought they were gospel.

Instead I would argue that one ought to use a different approach. Depending upon the discount rates and the assumptions I make about impacts, I can get methane indices that are a factor of two to four higher. Let's have a negotiation and agree internationally that we're going to use a number that's three times the GWP rather than just sort of pretending the emperor has clothes and use the GWP.

So my sense of the way you would use this method is as the basis for holding a discussion about the kinds of numbers that are logical, not as a sort of hardware tool that gets you the numbers. That would be silly.

Dr. Morgan: Dr. Nitze talked about technology transfer as a vehicle for helping developing countries out of the bind of allowing economic growth while simultaneously trying to constrain carbon dioxide emissions. One of the issues that arises, of course, in a place like China is that they are extraordinarily constrained in terms of generating capacity.

So while it's certainly true that the marginal cost of carbon emissions in China is a whole lot lower than it is here, if I help the Chinese with technology transfer of fluidized bed coal combustors or something like that, what's to prevent the net effect over the next few decades, of the Chinese installing more total capacity, not retiring any of the older stuff, but because they're terribly constrained, keep it on line as well. Because I'm providing a subsidy, I manage to accelerate the rate at which they can develop, which from an economic development point of view is fine, but from a CO_2 abatement point of view may have, at least for time scales of a few decades, the reverse impact of what I had hoped to achieve?

Dr. Nitze: Well, I think what you've left out is waste. The Chinese economy — and this is recognized throughout China — is still riddled with waste. They use probably twice as much primary energy per unit of actually delivered energy services as they would if they had the best current technology. There are numerous examples of that. If you take the coal cycle, they've got seven million people employed in a mining industry that produces slightly more coal than we do here, and we produce it with 140,000 miners. They transport that coal on railroads that I think have several gauges for the coal that still in some places has to be unloaded and reloaded.

Although coal subsidies have been substantially reduced, coal is not yet fully priced in China, at least at marginal replacement cost, and, therefore, there is an insufficient incentive for supply-side efficiency in electricity generation.

My latest information on the average conversion efficiency of a coal-fired boiler plant in China is something in the neighborhood of 20 percent. The comparable percentage for the United States is about 36 percent, and in Japan, there are not so many of them, but it's even higher. And so it goes right down the line. You go to the residential section of Beijing, and people are still burning coal in these little household furnaces, which are extremely inefficient energy converters and produce all of the particulates which you can breath if you're in Beijing in February or March. There aren't any efficient light bulbs, and for that matter, you know, the electric systems in most of their buildings could do with a substantial upgrade. There isn't sufficient insulation. There are no sophisticated energy management systems. I mean, you can go on and on.

Now, in that environment, there is a tremendous opportunity for basic efficiency improvements. And to their credit, the Chinese have been taking those opportunities for the last ten years or so, and energy intensity in China has been coming down at a rate of around three percent a year. They've done that with massive investments in things like district heating systems and combined cycle plans, where at least you use the excess heat to run another turbine.

But we have a generation or two to go in terms of further upgrades. And, again to their credit, they are very interested in what we've been trying to do in the United States in terms of, for example, our voluntary programs at EPA. China does want to create a Green Lights program. It's very interested in some of the other things that John Hoffman and his people have developed like Golden Carrot. We have a major CFC-free refrigerator program in China under way now.

I have argued: well, let's take the future step. Let's take the up-to-date Whirlpool technology, which is the result of our domestic competition, and introduce it into the smaller size refrigerators that are being manufactured and marketed in China.

If you factor in the demand side, what you hopefully could see, even without a major change in energy source, is that they could get double the energy service bang from the current pollution buck, and that carries you a long way. That buys you enough time to perhaps mitigate the risk of the tremendous increase in Chinese emissions while you are moving towards truly advanced technologies, such as fuel cells. There's no reason why China shouldn't be at the cutting edge of fuel cell research and why, in 10 or 20 years, whenever we solve some of the remaining problems, they shouldn't be using proton-exchange membrane fuel cells for a lot of their energy needs, both stationary and mobile.

I firmly believe that if we make them part of the ongoing technology revolution in the transportation and other sectors that we are undergoing, you will see a continued improvement, they will be able to substantially increase their quality of life to OECD levels with much less increase in pollution.

Ms. Werner: Green cars are likely to be expensive. They may need government subsidies to enter the U.S. market. Will similar subsidies be required in development countries?

Dr. Nitze: The short answer to that is yes, but there's an important qualifier. It's amazing what industry can do all over the world when presented with a firm public policy challenge. I will refer back to my meeting yesterday afternoon with people representing automotive manufacturers. Toyota is taking great pride in gradually increasing the size and weight of their vehicles, because we like heavier, more luxurious vehicles in the United States, but the fuel efficiency has remained the same. So, in other words, the Japanese have compensated, or at least Toyota has compensated, for the increase in size and features of their vehicles by steady improvements in efficiency. We have not done the same with our luxury vehicles, although they have not, to my knowledge, deteriorated in efficiency.

There's an article in the current Scientific American which demonstrates how much more progress we could make with vehicles just with current technologies. And yes, there is a front-end cost to doing that, but there are clever ways of reducing that front-end cost, and I rather like the idea of the Golden Carrot-type competition.

We have another possibility to use with the Ozone Transport Commission decision. The auto industry does not want to be locked into the California model of ZEVs and LEVs with fixed percentages by fixed dates nationwide. In return for a little flexibility, I think they may make a very significant commitment over a period of years to produce vehicles with much lower emissions profiles.

Now, I rather prefer the economic incentive route to the regulatory incentive route, but a combination of both, I think, can achieve results at much lower costs than it looks like on the front end.

I had to sign, at the Department of State, the economic impact statement for the Montreal Protocol, and it contained some really rather large numbers for U.S. compliance with the original, rather limited, phasedown that was required, the 20 and 50 percent, if you remember the '87 protocol in its initial form. And it was $3 billion or something like that. Our estimate, with hindsight, turned out to be nuts. We were going to have to spend a few extra bucks with respect to air conditioning, because CFC 22 is not going to be so easy to replace in the vehicle fleet. But beyond that, we've actually saved money with most of these substitutes.

And so I have a certain naive confidence that there are many other such opportunities out there.

Ms. Werner: What are chances for a coordinated strategy to promote diffusion and dissemination of renewables?

Dr. Nitze: I would say good, but one has to think that through. What you have to be careful about with renewables is prematurely backing winners. Let me give you an example with DOE's battery program. Phosphoric acid fluid cells are an improvement over lead acid batteries and other similar type systems but they're not the cutting edge by any means: we've made a big investment in that technology. You have to be careful that you don't create a rice bowl around a specific technology and international rice bowls are even worse than national rice bowls. They've become tougher to break. Some of the resource consortiums in Europe have build up tremendous institutional momentum and a lot of bureaucrats and people in industry have a stake. It then becomes difficult to change course.

So you need to collaborate. You certainly need international collaboration in providing institutional knowledge, in providing incentives, in providing ways in which renewable providers have an opportunity to offer their services in a cost-competitive way without necessarily betting on a particular technology, because the technologies are improving as we speak. And those choices, in my view, should be left, to the extent possible, to the marketplace.

Dr. Pershing: The only thing that I would add to that is to say that there is actually a fairly strong push from many of the OECD countries to look for ways that you can invest in technologies through collaborative activities. There is a proposal that has currently been tabled by the New Zealand delegation, which looks at how you might approach multi-nationals to have them coordinate their activities. The Swiss proposal and the German proposal both reflect some desire to see coordination of technological development.

But I would strongly agree, and I think that largely the international community would agree, that you ought to shy away from this picking of technologies. There is some concern that almost the only thing that you can say about the picking is that you probably have picked the wrong one.

Ms. Werner: Do examples of unsustainable development in North America and Eastern Europe, for example, burning rivers, disintegrating buildings, provide an adequate incentive for sustainable development in developing countries? How can this incentive, or percentage of an incentive, be advanced? Learn from our bad practices in a positive way.

Dr. Nitze: Unfortunately, there is a strong cultural element in the answer to that question. And, unfortunately, the cowboy culture is winning. One of the things people love about the American lifestyle is the freedom. You hop in your Bonneville and you drive to your Safeway and nobody tells you when and nobody tells you how, and the food is cheap and the videos are cheap and everything else is cheap and people love it.

I was with the chief Indian negotiator during the INC process, who's now their ambassador to China, and we had dinner together. And after excoriating the United States for its self-indulgence and a lack of discipline and over-consumption and so forth, he admitted that one of the great heroes on Indian television was J.R. Ewing from "Dallas," which was an extremely popular TV program. And if you've looked at the consumption patterns of the Indian middle class, you can see why. They like to live like a rich American family in Dallas.

In looking at Asia, I'm not sure that's going to be very easy to change. I think what we have to do is to get people to see it as in their self-interest to improve their quality of life in a somewhat different way. We've had that kind of revolution with respect to smoking with the middle or particularly upper middle class in this country. It is possible to get people to look at their quality of life in a different way through education and experience.

I think that's what we're going to have to do. People are not going to just leap at managed communities and parks with sort of interesting looking modular units limited to 900 square feet with a tie up with some little mini-bus that takes them to the local subway depot. Some of that is highly desirable. I think worldwide people are going to want to make their own choices. They're going to want to set their own agendas, and they're going to have to be provided with the tools to do that. And that, in a sense, is the problem.

Ms. Werner: Would you speculate on how emission targets for the so-called more advanced developing countries could be set?

Dr. Pershing: How they could be set, or how they could be received? My sense is that if you talk to the more advanced developing countries — and I think the kind of places you're talking about are perhaps Thailand and Korea, some of the countries in the developing part of the Asia Pacific region — you get a very negative response as your first reaction. They say, "Who me? I'm developing still."

On the other hand, you also begin to get, I think, a willingness to acknowledge that there are advantages in being the first out of the starting blocks, that a lot of what we think of, when we think of taking on commitments, also incudes a kind of welcome, if you will, to the brotherhood of the OECD. There are a lot of advantages that accrue to countries that can move quickly along that development path. And the additional incentives that might be provided by signing on to some of these international commitments may help them move along that path.

I think that the new frontier in competitive technologies is going in the direction of environmentally sound technologies. The willingness of countries to assume commitments from these newly industrialized countries and from the leaders of the developing country crowd will be in part contingent on their perception of the economic advantage to do so. And I think that that perception is changing, and it's changing in line towards a greater willingness to accept commitments.

Dr. Morgan: Although I might add that local and regional short-term environmental impacts are inevitably going to loom much larger than these sort of esoteric long-term things in the thinking of environmental agencies in these countries, and appropriately so.

Dr. Pershing: Because I think Granger is right. You're not going to sell this stuff any time in the near term on the global change impacts.

Ms. Werner: The Country Studies Program is having difficulty getting funding for FY '95. Won't it be even more difficult in this next year to obtain funding? Also, EPA's entire Office of International Activity seems to be vulnerable with the new Congress. What is the relative priority of climate change compared to other issues, for example, NAFTA implementation and the sidebar agreements?"

Dr. Nitze: Well, I'm going to answer those questions in reverse order. Actually, I have not played a very large role in climate change to date, but that's been largely a factor of the need to begin to get on top of other parts of what my office has to do.

I would say probably the most critical area that we're wrestling with is NAFTA implementation and our two borders. I would say the second very high geographic priority was Russia and what we're trying to do in the NIS and Eastern Europe. And the third is some of the major developing countries, particularly now in Latin America because of the interest in extending the free trade arrangements we have with Mexico to other Latin American countries and the concern about doing that in a way which factors in the environment.

There are also a number of global issues like trade and the environment, like land-based sources of pollution, where we're hosting a major conference which competes for attention. And historically, at least in recent years, the Policy Office and the Air Office have had the lead on climate change, and the International Offices have played a secondary role. But that may change slightly. I'm not going to be too aggressive about it, but over time, I would hope to have a somewhat larger voice.

Now, climate change funding for EPA, not specifically funding for my office but funding particularly for the policy office, and some of the research work, is somewhat vulnerable. Frankly, I would be surprised if it could be maintained at current levels.

So we're going to have to be quite clever about what we do with the money that is appropriated. I think some will be appropriated. I think the fears of drastic butchery by the new Congress are probably overwrought, but we will have to be clever.

The Countries Studies Program is a program for which a very strong case can be made on cost effectiveness grounds. For very little money, we get a lot of leverage on what other countries do, and that leverage can be translated into things like export opportunities for American firms and various other benefits to U.S. citizens. We have to make that case, but I think the case can be made and the money can be found.

With respect to OIA as a whole on a larger scale, we are going to have to make the case that what we do benefits U.S. citizens. There are two areas where I think that's pretty clear, the two borders with our neighbors, Mexico and Canada. There is another promising area, the promotion of exports of environmental goods and services as part of the Environmental Technology Initiative.

And frankly, I think there's a broader case on issues like climate change, which should appeal even to more conservative members of Congress. People are going to look to the U.S. to take the lead in responding to these global environmental challenges. It is in our interest to find clever ways of spreading the load and to get other countries to pick up a major share of that load, particularly the major developing countries. And the only way we can do that is through solid international diplomacy and participation in these multilateral processes. Frankly, that case has not been made as articulately as it should be, but I hope to help in that regard.

Ms. Werner: One additional point I might make about the Countries Studies Program is that it, in the past certainly, has received very much support from both sides of the aisle. And again, I quite agree that it has been a very, very effective, well-received program internationally. It has attracted great interest from the Global Environmental Facility in terms of its great value and ability to work with countries.

And another point I might also leave you with is the questions that we have heard raised in the Congress in the past with regard to actions that the United States should take with regard to climate change. These have often been coupled with: "why should we do things if other countries are not?" In terms of looking at where emissions are really going to be growing, Congress does not feel that we should be taking extra actions if other countries are not coming into the process.

This program is very clearly a way to encourage more and more countries to be willing and able to come into that international process as full partners. So I think that there is a very, very important case that can be articulated for the program, and it is extremely low cost for very great effects.

PROGRAM OF THE TENTH ANNUAL CONFERENCE OF THE CENTER FOR ENVIRONMENTAL INFORMATION, INC.

NOVEMBER 30, 1994

Session I: Global Climate Change Status

Chairman - **Robert Watson**, Deputy Director, White House Office of Science and Technology Policy

International Negotiating Committee Process
Oscar Avalle, Secretary of the Argentine Mission to the United Nations

U.S. Perspective
Daniel Reifsnyder, Office of Global Change, Department of State

Scientific Status Report and Discussion
Robert Watson, Deputy Director, White House Office of Science and Technology Policy

U.S. Climate Action Plan: An Overview
Karl Hausker, Deputy Assistant Administrator, Office of Policy, Planning, and Evaluation, U.S. Environmental Protection Agency

Session II: Perspectives on the U.S. Climate Action Plan

Chairman - **Nelson Hay**, Vice President, American Gas Association

State Perspective
David Rinebolt, National Association of State Energy Officials

Environmental Perspective
Daniel Lashof, Natural Resources Defense Council

Industry/Business Perspectives
John Hemphill, Business Council for a Sustainable Energy Future
Robert Beck, Edison Electric Institute

Critique of Session II
Elizabeth Cook, World Resources Institute
W. David Montgomery, Charles River Associates

DECEMBER 1, 1994

Session III: U.S. Plan - Mitigation Actions

Chairman - **Peter Fox-Penner**, Principal Deputy Assistant Secretary, Energy Efficiency and Renewable Energy, U.S. Department of Energy

Energy Supply
Charles Siebenthal, Electric Power Research Institute

Industrial Demand: The Climate Wise Program
Pamela Herman, U.S. Environmental Protection Agency

Electric Utilities: The Climate Challenge Program
Dirk Forrister, U.S. Department of Energy

Transportation
Wesley Warren, White House Office of Environmental Policy

Methane and Other Gases
Kathleen Hogan, U.S. Environmental Protection Agency

Funding the U.S. Climate Action Plan
T.J. Glauthier, Associate Director for Natural Resources, Energy and Science, EOP, Office of Management and Budget

Session IV: National Action Plans

Chairman - **David Jhirad**, Senior Energy Advisor, U.S. Agency for International Development

Canada
Richard Findlay, Environment Canada

Germany
Wolfram Schoett, Embassy of Germany

Japan
Satoshi Tanaka, Embassy of Japan

United Kingdom
Mark Hammond, Embassy of the United Kingdom

Asia
John Topping, Climate Institute

Critique of Session IV
Jennifer Morgan, U.S. Climate Action Network

DECEMBER 2, 1994

Session V: National Plans and the Climate Change Convention

Chairwoman - **Carol Werner**, Director, Energy Program, Environmental and Energy Study Institute

Integrated Assessment - A Tool For Policymaking
M. Granger Morgan, Department of Engineering & Public Policy, Carnegie-Mellon University

Innovative Responses
Irving Mintzer, Center for Global Change, University of Maryland at College Park

Setting Goals Under the Convention
Jonathan Pershing, Office of Global Change, Department of State

Closing Remarks
William Nitze, Assistant Administrator for International Activities, U.S. Environmental Protection Agency

PARTICIPANTS

Salud Astruc
Old Dominion Electric Cooperative
Glen Allen, VA

Oscar Avalle
Argentine Mission to the United Nations
New York, NY

Donna Banks
American Gas Association
Arlington, VA

Jeff Barber
McGraw-Hill Inside Energy
Washington, DC

Jack Barkenbus
University of Tennessee
Knoxville, TN

Robert Beck
Edison Electric Institute
Washington, DC

Chris Bernabo
Science & Policy Associates, Inc.
Washington, DC

Sharron Brown
NCI Information Services
McLean, VA

Hal Cardwell
Oak Ridge National Laboratory
Oak Ridge, TN

Jennifer Casey
Association of Home Appliance Manufacturers
Washington, DC

Edgar Chase
Clinton, MD

Elizabeth Cook
World Resources Institute
Washington, DC

James Crawford
American Standard Inc.
Arlington, VA

Cynthia Cummis
U.S. Environmental Protection Agency
Washington, DC

Patrick Curran
Allied Signal Inc.
Parsippany, NJ

David Downie
Columbia University
New York, NY

Harry Dowsett
Fairfax, VA

John J. Easton, Jr.
Edison Electric Institute
Washington, DC

David Feldman
University of Tennessee
Knoxville, TN

Steve Fetter
University of Maryland
Hyattsville, MD

Richard Findlay
Environment Canada
Ottawa, Ontario, Canada

Dirk Forrister
U.S. Department of Energy
Washington, DC

Peter Fox-Penner
U.S. Department of Energy
Washington, DC

Stacey Fried
Eastman Chemical Company
Washington, DC

Luc Gagnon
Hydro-Quebec
Montreal, Quebec, Canada

Robert Gehri
Southern Company Services, Inc.
Birmingham, AL

Paul Geise
St. Anne's - Belfield
Charlottesville, VA

Alfonso Gittienez
IWP's Clean Air Report
Arlington, VA

T.J. Glauthier
Office of Management and Budget
Washington, DC

Laura Gottsman
U.S. Environmental Protection Agency
Washington, DC

Mark Hammond
Embassy of the United Kingdom and Northern Ireland
Washington, DC

Sue Postle Hammond
Science & Policy Associates, Inc.
Washington, DC

David Hatherton
Water Furnace International
Fort Wayne, IN

Karl Hausker
U.S. Environmental Protection Agency
Washington, DC

Nelson Hay
American Gas Association
Washington, DC

Cecilia Hellstadius
Technical Outlook
Washington, DC

John Hemphill
Business Council for a Sustainable Energy Future
Washington, DC

Pam Herman
U.S. Environmental Protection Agency
Washington, DC

Carl Hocevar
RCG/Hagler Bailly, Inc.
Arlington, VA

Kathleen Hogan
Global Change Division
U.S. Environmental Protection Agency
Washington, DC

John Holt
NRECA
Washington, DC

Linda Horveth
Advocacy Institute
Washington, DC

Jau-Inn Huang
Excel Consulting Inc.
Wujih, Taichung, Taiwan

Vinke Jacoune
Paris, France

David Jhirad
U.S. Agency for International Development
Washington, DC

James Jones
Inside EPA
Arlington, VA

Joan M. Jordan
Private Consultant
Washington, DC

Anthony L. Joseph
New York State Energy Office
Albany, NY

John Justus
Science Policy Research Division
Library of Congress
Washington, DC

Ted Kennedy
DynCorp-Meridian
Alexandria, VA

Thomas Kirlin
American Petroleum Institute
Washington, DC

Wilfrid L. Kohl
Johns Hopkins University
Washington, DC

Daniel Lashof
Natural Resources Defense Council
Washington, DC

Sarah Laskin
World Wildlife Fund
Washington, DC

Othalene Lawrence
U.S. Department of Energy
Washington, DC

James Lunan
Peabody Holding Company, Inc.
St. Louis, MO

Robert Means
USI, Inc.
Washington, DC

Michael Mernick
U.S. Department of Energy
Boston, MA

Ann R. Mesnikoff
Center for Auto Safety
Washington, DC

Irving Mintzer
University of Maryland
College Park, MD

Nicola Mishev
Embassy of Bulgaria
Washington, DC

W. David Montgomery
Charles River Associates, Inc.
Washington, DC

Linda Moodie
National Oceanic and Atmospheric Administration
Washington, DC

Jennifer Morgan
US Climate Action Network
Washington, DC

M. Granger Morgan
Carnegie-Mellon University
Pittsburgh, PA

Wayne Morrissey
Congressional Research Service
Washington, DC

Valerie Newman
University of Delaware
Newark, DE

William Nitze
U.S. Environmental Protection Agency
Washington, DC

John Noller
Missouri Division of Energy
Jefferson City, MO

John J. Novak
Edison Electric Institute
Washington, DC

James O'Hear III
The Catholic University of America
Washington, DC

Judith Peatross
University of Virginia
Charlottesville, VA

Jonathan Pershing
U.S. Department of State - Office of Global Change
Washington, DC

Wendy Petry
CEI
Rochester, NY

Rickey Petty
U.S. Department of Energy
Washington, DC

David Pixler
UMUC
Rockville, MD

Elizabeth Pixley
Monroe Community College
Rochester, NY

Pam Pontillo
U.S. Department of Energy
Germantown, MD

Robert Pratt
Global Climate Change Digest
Valley Falls, NY

William Prendergast
ARI
Arlington, VA

Belinda Raband
Washington International Energy Group
Washington, DC

Wendy Ramsey
Oak Ridge National Laboratory
Oak Ridge, TN

Roberta Rand
U.S. Department of Agriculture
Beltsville, MD

Daniel Reifsnyder
U.S. Department of State
Washington, DC

Mike Reitz
U.S. Department of Energy
Washington, DC

Ken Richards
Battelle
Washington, DC

David Rinebolt
National Association of State Energy Officials
Washington, DC

Eric Rodenburg
World Resources Institute
Washington, DC

Jens Rosebrock
World Bank
Washington, DC

Bert W. Rust
NIST
Gaithersburg, MD

Charles Samuels
Association of Home Appliance Manufacturers
Washington, DC

Torrey Schobel
STAPPA/ALAPCO
Washington, DC

Wolfram Schoett
German Embassy
Office of Science and Technology
Washington, DC

B. Yvonne Selbig
Rochester Gas and Electric Corporation
Rochester, NY

Shirley Sherman
CEI
Rochester, NY

Rick Sherman
CEI
Rochester, NY

John Shlaes
Global Climate Coalition
Washington, DC

Charles Siebenthal
Electric Power Research Institute
Palo Alto, CA

Owawan Siriratpiriya
Chulalongkorn University
Thailand

Elizabeth Smith
Waterford, VA

David South
Argonne National Laboratory
Argonne, IL

Eugene Spejewski
Oak Ridge Institute for Science and Education
Oak Ridge, TN

Julian Roy Spradley, Jr.
Counselor on Environmental Law
Falls Church, VA

Fred Stoss
Oak Ridge National Laboratory
Oak Ridge, TN

Nicholas Sundt
Energy, Economics, and Climate Change
Washington, DC

Satoshi Tanaka
Embassy of Japan
Washington, DC

Ed Tatum
Old Dominion Electric Cooperative
Glen Allen, VA

James G. Taylor
Mobil Corporation
Fairfax, VA

Liz Thorndike
CEI
Rochester, NY

John Topping
Climate Institute
Washington, DC

Jim Turnure
U.S. Environmental Protection Agency
Takoma Park, MD

Rob Tyson
American Gas Association
Arlington, VA

Kristi Vanangu
Natural Resources Canada
Ottawa, Ontario, Canada

Angela VanDerwerken
Charles River Associates
Washington, DC

John Vitka, Jr.
Sandra National Labs
Livermore, CA

Larry D. Voorhees
Martin Marietta
Oak Ridge, TN

William Wagner
CEI
Rochester, NY

Wesley Warren
White House Office on Environmental Policy
Washington, DC

Robert Watson
White House Office of Science and Technology Policy
Washington, DC

Carol Werner
Environmental and Energy Study Institute
Washington, DC

James White
CEI
Rochester, NY

Robert Williams
Woodbridge, VA

Lisa Wolfe
American Forest and Paper Association
Washington, DC

Porter Womeldorff
Illinois Power Company
Decatur, IL

Barry Worthington
US Energy Association
Washington, DC

Ren-Tseng Young
ITRI, Energy, and Resources Lab
Chutung, Hsinchu, Taiwan

Pamela Zurer
Chemical & Engineering News
Washington, DC

INDEX

MIX
Papier aus verantwortungsvollen Quellen
Paper from responsible sources
FSC® C105338

If you have any concerns about our products,
you can contact us on
ProductSafety@springernature.com

In case Publisher is established outside the EU,
the EU authorized representative is:
Springer Nature Customer Service Center GmbH
Europaplatz 3, 69115 Heidelberg, Germany

Printed by Libri Plureos GmbH
in Hamburg, Germany